T0248741

BIOLOGICALLY ACTIVE NATURAL PRODUCTS

Microbial Technologies and Phyto-Pharmaceuticals in Drug Development

BIOLOGICALLY ACTIVE NATURAL PRODUCTS

Microbial Technologies and
Phyto-Pharmaceuticals in Drug Development

Edited by
Debarshi Kar Mahapatra, PhD
Swati Gokul Talele, PhD
Tatiana G. Volova, DSc
A. K. Haghi, PhD

First edition published 2021

Apple Academic Press Inc.
1265 Goldenrod Circle, NE,
Palm Bay, FL 32905 USA

4164 Lakeshore Road, Burlington,
.ON, L7L 1A4 Canada

CRC Press
6000 Broken Sound Parkway NW,
Suite 300, Boca Raton, FL 33487-2742 USA

2 Park Square, Milton Park,
Abingdon, Oxon, OX14 4RN UK

First issued in paperback 2021

© 2021 Apple Academic Press, Inc.

Apple Academic Press exclusively co-publishes with CRC Press, an imprint of Taylor & Francis Group, LLC

Cover images:
Top left: *Bidens torta,* the corkscrewbeggarticks. Eric Guinthe. https://commons.wikimedia.org/w/index.php?curid=2494037
Top right: *Aloe vera.* Photo by Erin Silversmith. https://commons.wikimedia.org/w/index.php?curid=475278
Bottom left: Neem (*Azadirachta indica*) in Hyderabad. Photo by J.M. Garg. https://commons.wikimedia.org/w/index.php?curid=7069304
Bottom right: *Abutilon indicum var.* indicum. Photo by Bō-á-tún_ê_hoe. https://commons.wikimedia.org/w/index.php?curid=820757

Library and Archives Canada Cataloguing in Publication

Title: Biologically active natural products : microbial technologies and phyto-pharmaceuticals in drug development / edited by Debarshi Kar Mahapatra, PhD, Swati Gokul Talele, PhD, Tatiana G. Volova, DSc, A.K. Haghi, PhD.

Names: Mahapatra, Debarshi Kar, editor. | Talele, Swati Gokul, editor. | Volova, Tatiana G., editor. | Haghi, A. K., editor.

Description: Includes bibliographical references and index.

Identifiers: Canadiana (print) 20200327720 | Canadiana (ebook) 20200327887 | ISBN 9781771889049 (hardcover) | ISBN 9781003057505 (hardcover)

Subjects: LCSH: Drug development. | LCSH: Natural products. | LCSH: Biological products. | LCSH: Microbial biotechnology. | LCSH: Materia medica, Vegetable. | LCSH: Nanotechnology.

Classification: LCC RM301.25 .B56 2021 | DDC 615.1—dc23

Library of Congress Cataloging-in-Publication Data

CIP data on file with US Library of Congress

ISBN: 978-1-77188-904-9 (hbk)
ISBN: 978-1-77463-927-6 (pbk)
ISBN: 978-1-00305-750-5 (ebk)

About the Editors

Debarshi Kar Mahapatra, PhD
Assistant Professor, Department of Pharmaceutical Chemistry, Dadasaheb Balpande College of Pharmacy, Rashtrasant Tukadoji Maharaj Nagpur University, Nagpur, Maharashtra, India

Debarshi Kar Mahapatra, PhD, is currently an Assistant Professor in the Department of Pharmaceutical Chemistry at Dadasaheb Balpande College of Pharmacy, Rashtrasant Tukadoji Maharaj Nagpur University, Nagpur, Maharashtra, India. He was formerly Assistant Professor in the Department of Pharmaceutical Chemistry, Kamla Nehru College of Pharmacy; RTM Nagpur University, Nagpur, India. He has taught medicinal and computational chemistry at both the undergraduate and postgraduate levels and has mentored students in their various research projects. His area of interest includes computer-assisted rational designing and synthesis of low molecular weight ligands against druggable targets, drug delivery systems, and optimization of unconventional formulations. He has published research, book chapters, reviews, and case studies in various reputed journals and has presented his work at several international platforms, for which he has received several awards by a number of bodies. He has also authored the book titled Drug Design. Presently, he is serving as a reviewer and editorial board member for several journals of international repute. He is a member of a number of professional and scientific societies, such as the International Society for Infectious Diseases (ISID), the International Science Congress Association (ISCA), and ISEI.

Swati Gokul Talele, PhD
Assistant Professor, Department of Pharmaceutics, Sandip Institute of Pharmaceutical Sciences, Savitribai Phule Pune University, Pune, Maharashtra, India

Swati Gokul Talele, PhD, is currently serving as an Assistant Professor, Department of Pharmaceutics at Sandip Institute of Pharmaceutical Sciences. She has 18 years of experience in research along with teaching. She has published more than 20 research papers in various reputed international and national journals as well as more than 30 review papers. She has also presented research work at several conferences and has received several awards. She has authored the book titled *Natural Excipients* and has many chapters and books that are in progress. She has supervised many MPharm students and is currently associated with many research projects. She also worked as a College Examination Officer (CEO) for more than three years and is a member of the university examination committee. Dr. Talele has delivered interactive talks at continuous education programs for registered pharmacists and is a life member of the Association of Pharmacy Teachers of India (APTI) and a member of the Indian Pharmaceutical Congress Association (IPCA). Her interests are in the field of nanotechnology, natural polymers, herbal formulations, and radiolabeling-based bio-distribution studies.

Tatiana G. Volova, PhD
Professor and Head, Department of Biotechnology, Siberian Federal University, Krasnoyarsk, Russia

Tatiana G. Volova, DSc, is a Professor and Head, Department of Biotechnology at Siberian Federal University, Krasnoyarsk, Russia. She is the creator and head of the Laboratory of Chemoautotrophic Biosynthesis at the Institute of Biophysics, Siberian Branch of the Russian Academy of Sciences. Professor Volova is conducting research in the field of physico-chemical biology and biotechnology and is a well-known expert in the field of microbial physiology and biotechnology. Dr. Volova has created and developed a new and original branch in chemoautotrophic biosynthesis, in which the two main directions of the XXI century technologies are conjugate, hydrogen energy, and biotechnology. The obtained fundamental results provided significant outputs and were developed by the unique biotechnical producing systems, based on hydrogen biosynthesis for single-cell protein, amino acids, and enzymes. Under the guidance of Professor Volova, the pilot production facility of single-cell protein (SCP), utilizing hydrogen, had been created and put into operation. The possibility of involvement of man-made sources of hydrogen into biotechnological processes as a substrate, including synthesis gas from brown coals and vegetable wastes, was demonstrated in her research. She had initiated and deployed in Russia the comprehensive research on microbial degradable bioplastics; the results of this research cover various aspects of biosynthesis, metabolism, physiological role, structure, and properties of these biopolymers and polyhydroxyalkanoates (PHAs), and have made a scientific basis for their biomedical applications and allowed them to be used for biomedical research. Professor Volova is the author of more than 300 scientific works, including 13 monographs, 16 inventions, and a series of textbooks for universities.

A. K. Haghi

Professor Emeritus of Engineering Sciences, Former Editor-in-Chief, International Journal of Chemoinformatics and Chemical Engineering and Polymers Research Journal; Member, Canadian Research and Development Center of Sciences and Culture

A. K. Haghi, PhD, is the author and editor of 200 books, as well as 1000 published papers in various journals and conference proceedings. Dr. Haghi has received several grants, consulted for a number of major corporations, and is a frequent speaker to national and international audiences. Since 1983, he served as a professor at several universities. He is former Editor-in-Chief of the *International Journal of Chemoinformatics and Chemical Engineering* and *Polymers Research Journal* and is on the editorial boards of many international journals. He is also a member of the Canadian Research and Development Center of Sciences and Cultures (CRDCSC), Montreal, Quebec, Canada.

Contents

Contributors

Eknath Ahire
Divine College of Pharmacy, Nanpur Road, Satana, Nasik, Maharashtra, India

Aishwarya A. Andhare
Department of Microbiology, Biotechnology, and Chemistry, Dayanand Science College,
Latur – 413512, Maharashtra, India

Meghawati R. Badwar
Department of Pharmaceutics, Sandip Institute of Pharmaceutical Sciences, Nashik Maharashtra India

Akshada A. Bakliwal
Department of Pharmaceutics, Sandip Institute of Pharmaceutical Sciences, Nashik Maharashtra India

Sanjay Kumar Bharti
Institute of Pharmaceutical Sciences, Guru Ghasidas Vishwavidyalaya (A Central University),
Bilaspur – 440037, Chhattisgarh, India

Anshda Bhatnagar
Department of Health and Life Sciences, Coventry University, Coventry CV15FB, United Kingdom

Shilpa Borkar
Department of Pharmacology, Kamla Nehru College of Pharmacy, Nagpur – 441108,
Maharashtra, India

Gloria Castellano
Departamento de Ciencias Experimentales y Matemáticas, Facultad de Veterinaria y Ciencias
Experimentales, Universidad Católica de Valencia San Vicente Mártir, Guillem de Castro-94,
E-46001 València, Spain

Kirti Dubli
Hislop School of Biotechnology, Hislop College, Nagpur – 440001, Maharashtra, India,
E-mail: kirtidubli@gmail.com

Evgeniy Gennadievich Kiselev
Siberian Federal University, 79 Svobodnyi Av., Krasnoyarsk – 660041, Russia;
Institute of Biophysics SB RAS, Federal Research Center "Krasnoyarsk Science Center SB RAS,"
50/50 Akademgorodok, Krasnoyarsk – 660036, Russia

Shailaja Latkar
Department of Pharmacology, Kamla Nehru College of Pharmacy, Nagpur – 441108, Maharashtra,
India

Debarshi Kar Mahapatra
Assistant Professor, Department of Pharmaceutical Chemistry, Dadasaheb Balpande College of
Pharmacy, Nagpur – 440037, Maharashtra, India; Department of Pharmaceutics,
Gurunanak College of Pharmacy and Technical Institute, Nagpur – 440026, Maharashtra India,
E-mails: dkmbsp@gmail.com; mahapatradebarshi@gmail.com

Swapnali A. Patil
Department of Pharmaceutics, Sandip Institute of Pharmaceutical Sciences, Nashik Maharashtra India

Vaibhav Shende
Department of Pharmaceutics, Gurunanak College of Pharmacy and Technical Institute,
Nagpur – 440026, Maharashtra India

Ravindra S. Shinde
Department of Microbiology, Biotechnology, and Chemistry, Dayanand Science College,
Latur – 413512, Maharashtra, India, E-mail: rss.333@rediffmail.com

Ekaterina Igorevna Shishatskaya
Siberian Federal University, 79 Svobodnyi Av., Krasnoyarsk – 660041,
Russia; Institute of Biophysics SB RAS, Federal Research Center
"Krasnoyarsk Science Center SB RAS," 50/50 Akademgorodok, Krasnoyarsk – 660036, Russia

Gokul S. Talele
Matoshree College of Pharmacy, Eklahare, Nashik, Maharashtra, India

Swati G. Talele
Department of Pharmaceutics, Sandip Institute of Pharmaceutical Sciences, Mahiravani,
Nashik Maharashtra India, E-mail: swatitalele77@gmail.com

Francisco Torrens
Institut Universitari de Ciència Molecular, Universitat de València, Edifici d'Instituts de Paterna,
P. O. Box 22085, E – 46071 València, Spain, E-mail: torrens@uv.es

Tatiana Grigorievna Volova
Siberian Federal University, 79 Svobodnyi Av., Krasnoyarsk – 660041, Russia;
Institute of Biophysics SB RAS, Federal Research Center "Krasnoyarsk Science Center SB RAS,"
50/50 Akademgorodok, Krasnoyarsk – 660036, Russia, E-mail: volova45@mail.ru

Seema Wakodkar
Department of Pharmacology, Kamla Nehru College of Pharmacy, Nagpur – 441108,
Maharashtra, India, E-mail: seemausare@rediffmail.com

Abbreviations

3HHx	3-hydroxyhexanoate
3HO	3-hydroxyoctanoate
3HV	3-hydroxyvalerate
4-AP	4-amino pyridine
4-APIC	4-amino pyridine induced convulsion
4HB	4-hydroxybutyrate
ACTH	adrenocorticotrophic hormone
AIA	antigen-induced arthritis
ALP	alkaline phosphatase
ALT	alanine aminotransferase
AM	amphiphilic molecules
AOA	antioxidant activity
AP	activator protein
APIs	active pharmaceutical ingredients
AQ	assimilation quotient
Art	artemisinin
AST	aspartate aminotransferase
BCA	bicinchoninic acid
BHB	biomass of hydrogen bacteria
BL	butyrolactone
Ca^{2+}	calcium ions
CAT-1	cationic amino acid transporter-1
CBB	coomassie brilliant blue
CC	click chemistry
CCS	croscarmellose sodium
Cdks	cyclin-dependent kinases
CH_2O	carbohydrate
CHM	Chinese herbal medicine
CL	chemical ligation
CO	carbon monoxide
CO_2	carbon dioxide
COX	cyclo-oxygenase
CRI	chronic renal insufficiency

DAO	D-amino acid oxidase
DBil	direct bilirubin
DHAA	dihydroartemisinic acid
DM	diabetes mellitus
DO	dis-fathomed oxygen
ECM	extracellular matrix
EGF	epidermal growth factor
EP	ethnopharmacology
EPM	elevated plus maze
ER	endoplasmic reticulum
FDNB	1-fluoror-2, 4-dinitrobenzene
FSH	follicle-stimulating hormone
Fα	follitropin α
GABA	γ-aminobutyric acid
GHD	growth hormone deficiency
GHRH	growth hormone-releasing hormone
GI	gastrointestinal
GnRHa	GnRH agonists
GnRHag	GnRH antagonists
GRAS	generally recognized as safe
GSH	glutathione
HbA	adult hemoglobin
HbF	fetal hemoglobin
H-bonds	hydrogen bonds
HBP	hamster buccal pouch
HDL	high-density lipoproteins
HPMC	hydroxypropyl methylcellulose
HTS	high-throughput screening
IDL	intermediate-density lipoproteins
IGF-1	insulin growth factor-1
IGF-2	insulin growth factor-2
IGFs	insulin growth factors
IM	intramuscular
LDL	low-density lipoproteins
LDT	light-dark transition
LH	luteinizing hormone
LSS's	life support systems
MAPKs	mitogen-activated protein kinases
MB	metoxybutyrate

MBC	minimum bactericidal concentration
MCC	microcrystalline cellulose
MDA	malondialdehyde
MESIC	maximal electroshock-induced convulsion
MIC	minimum inhibitory concentration
MSG	monosodium glutamate
MSU	model of monosodium urate
NaOH	sodium hydroxide
NCL	native chemical ligation
NIH	National Institute of Health
NO	nitric oxide
NPs	natural products
OA	osteoarthritis
OFT	open-field test
PAAc	polycarbophilpolyacrylic acid
PDA	photodiode array
PEO-PPO-PEO	poly(ethylene oxide)-poly(propylene oxide)-poly(ethylene oxide)
PHAs	polyhydroxyalkanoates
PIC	picrotoxin-induced convulsion
PKB	protein kinase B
PMAA-gEG	poly(methacrylic acid g-ethylene glycol)
POMC	pro-opiomelanocortin
PTH	parathormone
PTZ	pentylenetetrazole
PTZIC	pentylenetetrazole induced convulsions
PWS	Prader-Willi syndrome
RA	rheumatoid arthritis
RANK	receptor activator of NF-κB
RCr	renal creatinine
RFLP	restriction fragment length polymorphism
ROS	reactive oxygen species
RQ	respiratory quotient
RU	urea
SAM	S-adenosyl methionine
SBP2	SECIS-binding protein 2
SC	subcutaneous
SCP	single-cell protein
SCr	serum creatinine

Se	selenium
Sec	selenocysteine
SelA	selenocysteine synthetase
SGA	small for gestational age
SHOX-D	short stature homeobox-containing gene deficiency
SIC	strychnine-induced convulsions
SL	Staudinger ligation
SMA	spontaneous motor activity
SPPS	solid-phase peptide synthesis
SPR	surface plasmon resonance
SPS	solution-phase synthesis
SPS2	seleno-phosphate synthetase 2
SRP	signal recognition particle
SSF	solid-state fermentation
SSG	sodium starch glycolate
ST	somatotrophin
STAT	signal transducer and activator of transcription
STL	sesquiterpene lactone
STR	ST receptors
TBil	total bilirubin
TEA	triethanolamine
TGF	transforming growth factor
TM	traditional medicine
TP	total proteins
TS	turner syndrome
UPLC	ultra-performance liquid chromatography
VLDL	very-low-density lipoproteins
WHO	World Health Organization
ZOI	zone of inhibition

Preface

Biologically active natural products (NPs) have provided considerable value to the pharmaceutical industry over the past half-century.

Natural products and their substructures have long been valuable starting points for medicinal chemistry and drug discovery. Since the earliest days of medicine, we've turned to nature for our treatments. "Natural medicines" (which include animal- and mineral-sourced medicines as well as plants) have been significant source materials for medicine discovery over the past several decades. Thousands of compounds have been isolated from them, developed into pharmaceuticals, and used as conventional medicines. Herbal supplements, nutraceuticals, and other herbal healthcare products are also becoming familiar in our daily lives.

In recent decades, natural products have undisputedly played a leading role in the development of novel medicines. Yet, trends in the pharmaceutical industry at the level of research investments indicate that natural product research is neither prioritized nor perceived as fruitful in drug discovery programs as compared with incremental structural modifications.

The tree, which is called Neem (*Azadirachta indica*) is a natural tropical evergreen tree and found to be deciduous in drier regions native to the Indian sub-continent. Chapter 1 focuses on the ethnopharmacological cum modern perspectives of Neem (leaf, seed, bark, flower, and oil) extracts and products in context to diabetes, cancer, hypertension, AIDS, infections, ulcer, fertility, and gynecological problems, common fever, sexually transmitted diseases, skin diseases, leprosy, and dental diseases. Comprehensive phytochemical structures are highlighted in this chapter.

In Chapter 2, it is shown that peptides and proteins are a major fraction of all the molecules in the body. They are everywhere in the body transmitting messages in the form hormones, quickening the reaction as enzymes, transport oxygen via hemoglobin like transport proteins, providing structural skeleton to the organisms in the form of collagen present in every type of connective tissue, in therapeutics as agonist and analog of peptides, etc. The resourcefulness of proteins is an essence of life.

Chapter 3 provides an overview of the results of a study of the biotechnological potential of hydrogen-oxidizing bacteria, obtained at the Institute of Biophysics SB RAS and Siberian Federal University. The controlled culture of hydrogen bacteria for several years has been studied in three directions: (i) the hydrogen bacteria-regenerative link of human life support systems (LSSs); (ii) Biomass of hydrogen bacteria (BHB)-potential protein source; (iii) Hydrogen bacteria and the synthesis of biodegradable polymers. It has been shown that a culture of hydrogen together with water electrolysis can solve the main tasks of human LSSs: oxygen supply, assimilation of released CO_2, utilization of human wastes and water treatment, and synthesis of protein-rich biomass. Technology has been developed for the synthesis of protein biomass on hydrogen bacteria (BHB); 10 tons of biomass were studied in rations of tipsy, farm animals, and fur-bearing animals. The possibility of replacing 35–50% of proteins in BHB feeds without a decrease in animal productivity was shown. The ability of hydrogen-oxidizing bacteria to synthesize PHA under autotrophic and heterotrophic conditions makes them good candidates for commercial production of PHAs. Researchers tested different modes of cultivation of hydrogen bacteria and on hydrogen, and synetzivroany PHAs of various chemical compositions. The influence of the set and ratio of monomers in PHA on their basic properties was studied. The prospects of the use of polymers for various applications.

Xanthomonas axonopodis pv. punicae causes bacterial blight disease in pomegranate. The present investigation was initiated to find a suitable alternative to synthetic antibiotics for the management of plant diseases caused by bacteria. Chapter 4 was aimed to use wild plant species viz., *Abutilon indicum, Prosopis juliflora,* and *Acacia arabica* as Antibacterial agent against *Xanthomonas axonopodis pv. punicae.* The aqueous extracts of *Abutilon indicum, Prosopis juliflora,* and *Acacia arabica* plants has Antibacterial activity against *Xanthomonas axonopodis pv. punicae.* The antibacterial activity was tested by well diffusion assay, minimum inhibitory concentration (MIC), and minimum bactericidal concentration (MBC). The maximum activity recorded in *P. juliflora* (MIC = 1.03 mg ml^{-1} and MBC = 0.15 mg ml^{-1}) and *A. arabica* (MIC = 1.00372 mg ml^{-1} and MBC = 2.58 mg ml^{-1}) against *X. axonopodis pv. Punicae*, while the lowest activity was recorded by *A. indicum* (MIC = 0.619 mg ml^{-1} and MBC = 0.923 mg ml^{-1}). The highest ZOI was shown by *P. juliflora* while lowest ZOI was shown by *A. indicum* The results infer that the extracts of *Prosopis juliflora* and *Acacia Arabica* are highly sensitive against the *Xanthomonas axonopodis pv. punicae.* Plant extracts exhibited antibacterial activity with the potential to be used in the management of many plant diseases as an alternative to chemical antibiotics. The further phytochemical analysis is required to identify the bioactive compounds responsible for antibacterial activity.

Microbial biotechnology includes the utilization, genetic control, and modification of micro-organisms for making commercially valuable products, which involve fermentation and various upstream and downstream processes. From the early years, microorganisms utilized for supplying products such as bread, beer, and wine. Microorganisms produce a stunning exhibit of profitable products that are necessary for their own advantages. Agricultural production is important to fulfill nourishment prerequisites for the developing total population. However, its acknowledgment is related with the mass utilization of non-inexhaustible characteristic assets and with the outflow of greenhouse gases causing atmosphere changes. The researcher's challenge is to meet practical ecological and conservative issues without trading off yields. This chapter includes the exploration endeavors that went for improving a reasonable and solid generation through a suitable administration. Also, it includes advancement in technologies for formulation and applications dependent on specific plant-related microorganisms. At last, the challenges and opportunities to oversee normally existing microbial populations, including those non-culturable, are investigated in Chapter 5.

In Chapter 6, it is shown that color has a complex artistic, physiological, symbolic, psychological, and associative role for humans. The modern researches have suggested that product color also influence the therapeutic efficacy. The chapter exclusively focuses on the role, applications, utility, safety guidelines (as per FDA, WHO, EC, etc.), environmental aspects, handling precautions, stability, blending, storage, regulatory concern (European Union Legislation, United States Union Legislation, Licensing Authority Approval, and The Food, Drug, and Cosmetic Act) of various synthetic and natural colorants, dyes, pigments, aluminum or calcium lakes, inorganic colors, etc., like β-carotene, indigo carmine lake, sunset yellow lake, brilliant blue lake, amaranth lake, quinoline yellow lake, Allura red lake, tartrazine, erythrosine, Patent Blue V, etc., in pharmaceutical dosage forms such as tablets (wet granulation and direct compression), pellet, tablet coating (sugar coating and film coating), capsules (hard gelatin capsule and soft gelatin capsule), ointment, solution, toothpaste, etc. In this modern avenue, the colorants are needed to be judged for their benefit ratio, regulatory restrictions, innovations, discoveries, and safety.

In Chapter 7, microbial pigments are obtained from pure natural origin, the origin is the renewable that is simply degradable, and without manufacturing of refractory intermediates when they arrive in the atmosphere. Natural pigments have increasing significance not only in colors but also since of their harmless and medicinal characteristics. The consciousness amongst societies towards natural pigments and there safe properties with very lesser side effects as compared to artificial colors and dyes. The uninterrupted use of artificial colorants or pigments not only causes environmental pollution but also lots of fitness-related complications in humans. Consequently, it is necessary to explore different natural origins of pharmaceutical and food-grade pigments and their prospective applications. In future natural products manufactured by microbes as pigments are nontoxic and healthier than artificial pigmented products.

In Chapter 8, it is shown that the plants are the leading source with therapeutic significance from thousands of years. *T. divaricata* L. (Apocynaceae) renowned medicinal plant used to treat a substantial number of human ailments and extensively scattered in the various parts of Africa, America, and Asia. The study critically evaluates the phytochemistry, pharmacology, ethnobotanical, and medicinal uses of *T. divaricata* L. species. The plant in whole or its peculiar segments applied for treatment against ulceration, fracture, postnatal recovery, syphilis, fever, tumors, and orchitis in form of boiled juice, decoctions infusions, and poultices, in the region of Southeast Asia. The various studies cataloged alkaloids as the influential phytochemical in accession to phenols, saponins, and sterols with immense biological activities such as antimicrobial, analgesic, anthelmintic, vasorelaxation, antiviral, and cytotoxicity. The medicinal uses of different parts of T. divaricate across Asia have been estimated by scientific data. Future studies focus on different phytoconstituents, such as jerantinine and vincamajicine alkaloids structure-bioactivity relationship, which could potentially improve the future application towards reversing anticancer drug resistance.

The aim of Chapter 9 is to concentrate on differing kinds of Aloe plant formulations and preparation procedures or methodology. In many parts of the world, there is a rich tradition of the oldest medicinal plant ever known and the most applied medicinal plant formulations. The Aloe plant has been renowned and in use for hundreds of years for its health, beauty, medicinal, and skincare properties. In the whole world, there are over 300 species of Aloe plant, which is succulent and the most common form is *Aloe barbadensis* Miller (Family: Liliaceae). It may be a woody plant or dendriform, perennial, xerophytic, succulent, or chromatic color plant with triangular, fleshy leaves with saw-toothed edges, yellow hollow flowers, and fruits that contain varied seeds. Multiple clinical studies have confirmed its treatment as it has anti-oxidant activity against free radicals, accelerates the healing of burns, reduces constipation, may improve skin and prevent wrinkles, lower blood sugar level, dentistry for wound healing effect, etc.

The leaves contain 75 active constituents such as vitamins, minerals, natural sugar, enzymes, lignin, saponins, hydroxy acid, amino acid, etc. The varied bioactive compounds exhibit emollient, purgative, medicinal drug, anti-microbial, anthelmintic, antifungal, aphrodisiac, anti-septic, and cosmetic values. The study of different formulations of *Aloe vera* plant like a gel, beads, tablet, capsule, nanoparticles, microparticles, face mask, hydrogel, film, emulgel, transdermal patch, nanosuspension, chewing gum, etc.

The aim of Chapter 10 is to develop an accurate and fast ultra-performance liquid chromatography (UPLC) coupled with a photodiode array (PDA) method for the simultaneous determination of artemisinin (Art), arteannuin B, arteannuin C, dihydroartemisinic acid (DHAA) and artemisinic acid in *Artemisia annua*. The established ultra-performance liquid chromatography-photodiode array method is valuable for improving the quantitative analysis of sesquiterpene components in *Artemisia annua*. Budlein A is a sesquiterpene lactone (STL) with antinociceptive and anti-inflammatory properties, related to the inhibition of pro-inflammatory cytokines and neutrophil recruitment. The effect of budlein A is evaluated in antigen-induced arthritis (AIA) in mice. Gout is the most common inflammatory arthritis worldwide. It is a painful inflammatory disease, induced by the deposition of monosodium urate crystals in the joints and peri-articular tissues. Sesquiterpene lactones are secondary metabolites biosynthesized mainly by species from the family Asteraceae. They present anti-inflammatory, analgesic, antitumoral, anti-parasitic, and antimicrobial activities. The efficacy of sesquiterpene lactone budlein A is evaluated in a model of acute gout arthritis in mice.

CHAPTER 1

Azadirachta indica: Imperative Mini-Opinions on an Ethnopharmacological Savior

ANSHDA BHATNAGAR[1] and DEBARSHI KAR MAHAPATRA[2]

[1]*Department of Health and Life Sciences, Coventry University, Coventry CV15FB, United Kingdom*

[2]*Department of Pharmaceutical Chemistry, Dadasaheb Balpande College of Pharmacy, Nagpur – 440037, Maharashtra, India*

1.1 INTRODUCTION

The tree which is called neem (*Azadirachta indica*) a natural tropical ever-green tree and found to be deciduous in drier regions native to the Indian subcontinent [1]. It has been found that neem is used extensively in Ayurvedic medicine for more than 1000's of years because of its long-lasting medicinal properties. Neem is regarded as *'arista'* in Sanskrit, this word means *'perfect, imperishable, and complete'* [2]. Most of the parts of this plant such as fruits, seeds, leaves, bark, and roots contain components, which are proved to have antiseptic, antiviral, antipyretic, anti-inflammatory, antiulcer, and antifungal uses. The Sanskrit name for neem called *'nimba'* taken from the term *'nimbatiswasthyamdadati'* which signifies *'to give good health.'* The significance of the Neem tree is scripted in ancient documents *'Charak-Samhita'* and *'Susruta-Samhita,'* which builds the basics of the Indian system of natural treatment, Ayurveda. It is generally known as 'Margosa' or 'Indian lilac' and associates to the family Meliaceae. The Persian name of neem is termed *'Azad-Darakth-E-Hind'* which denotes the 'Free tree of India.' Neem is accepted to be a division of India's genetic diversity [3, 4].

Neem tree is considered to be the most researched tree in the entire world and is regarded to be the most beneficial tree of the 21st century. It has been found that Neem holds great potential in the streams of environment protection, pest management, and medicine. Neem is a natural medium of pesticides, agrochemicals, and insecticides [5, 6].

Neem is a huge tree with a height, which grows up to 25 m, semi-erect to the straight trunk, has a girth of 3 m, and enlarged branches forming a broad ring. Generally, the Neem tree initially starts fruiting after a gap of 3–5 years. In a span of 10 years, this tree becomes fully mature to produce. After 10 years and onwards, this tree becomes capable of producing up to 50 kg of fruit annually [7].

The plant 'Neem' is reported to have a life of about two centuries. The tree has proved adaptability to a vast range of climatic, edaphic, and topographic conditions. It rebels well in stony, dry, and shallow soils. Also, on soils, which have a solid calcareous or clay pan, at a facile depth. Neem tree needs a small amount of water and a huge amount of sunlight [3, 4].

This tree nurtures naturally in regions where rainfall of range 450 to 1200 mm is experienced. However, it has been proved that even in areas where the rainfall is as less as 150 to 250 mm this tree can grow. Neem is noticed to grow on altitudes which extend up to the area of 1500 m [8–10]. This tree grows good in the vivid temperature range of 0°C to 49°C [11]. This tree is not prone. This tree withstands water-logged areas and a soil type which is poorly drained. It is found that this tree grows best in the pH, which ranges between 4 and 10. It is famous for its attribute that it can be grown on multiple types of soil including clayey, saline, and alkaline soil, but is noticed to be grown best on deep well-drained soil and black cotton soils with good partial soil water. Neem trees have the classic property to neutralize acidic soils by a notable property of calcium mining [11].

All these parts are proven to contain compounds which have antifungal, antiseptic, anti-inflammatory, antipyretic, and others use [12]. Traditionally, the Neem tree is used for the treatment of commonly known diseases like malaria, tuberculosis, leprosy, and fever. Extracts of Neem are traditionally administered through the routes which include its application-oral, topical uses, and vaginal [13].

1.2   GENERAL INFORMATION

1.2.1   TAXONOMY CIRCUMSCRIPTION [14]

- **Kingdom:**   Plantae
- **Order:**   Ratulae

- **Suborder:**   Ratinae
- **Family:**   Malaceae
- **Subfamily:**   Maleoideae
- **Tribe:**   Miliceae
- **Genus:**   Azadirachta
- **Species:**   Indica

1.2.2   BOTANICAL DESCRIPTION

This tree gains a height of about 35–50 ft and higher. This tree has branches that are massively spread and an erect trunk with dark and rough bark forms. Its bark has comprehensive fissures bifurcated by ridges. Fruits are green and turn yellow when ripe. These fruits have a pungent smell. The leaves are imparipinnate, compound, and each leaf consists of 5–15 leaflets. A leaf has multiple panicles which flower later on leaf's axil. These leaves produce drupes that are yellow in color and glabrous in structure. These drupes are 12–20 mm long in height. It is an evergreen tree but fresh and flowers come. In March–April, whereas fruits grow and mature in the months of July and August. The growth of fresh leaves and fruit depends on the locality and atmospheric conditions [15].

1.2.3   GEOGRAPHICAL DISTRIBUTION

Its maximal growth is found in South-East Asia and West Africa. They have been recently found out to grow best in Central America and the Caribbean. In Asian countries, mostly in India, it is observed that neem occurs naturally in Siwalik Hills which are dry forests of Andhra Pradesh. It is Cultivated and Naturalized all along the dry regions of tropical and sub-tropical.

1.2.4   PHYTOCHEMISTRY

Active principles which are extracted from the different sub-units of this neem tree are Nimbidin, Azadirachtin, Salanin, valassin, gedwin; get along to make up the bitter constituents of neem oil. 30–50% of oil which is taken from the kernels of neem is used by the pesticides, soap, pharmaceutical industries and contains many active ingredients which together are known as triterpene or liminoids. The top-four best known liminoid compounds include nimbin, meliantriol, salannin, and azadirachtin. A structural list of phytochemicals is provided which comprises of three primary classes: terpenes (Figure 1.1), non-terpenes (Figure 1.2), and miscellaneous (Figure 1.3) [16].

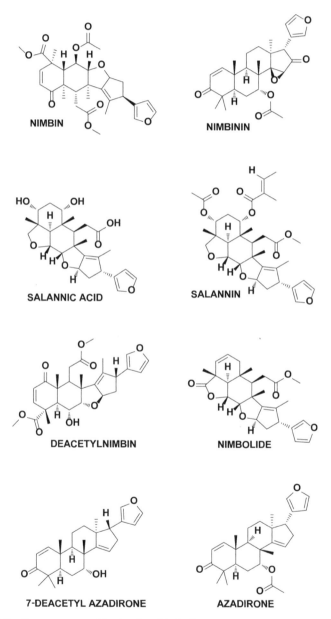

FIGURE 1.1   Terpenes present in *Azadirachta indica*.

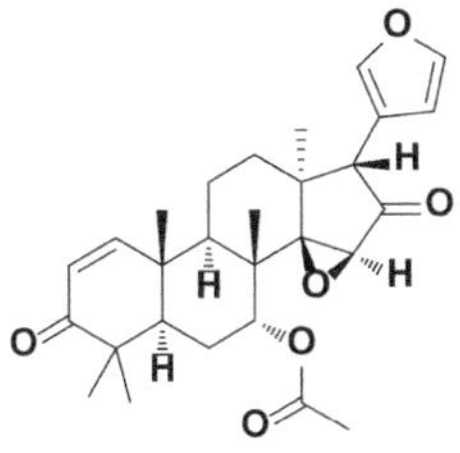

EPOXYAZADIRADIONE

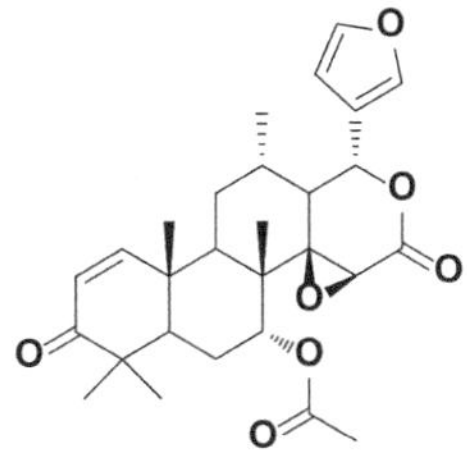

GEDUNIN

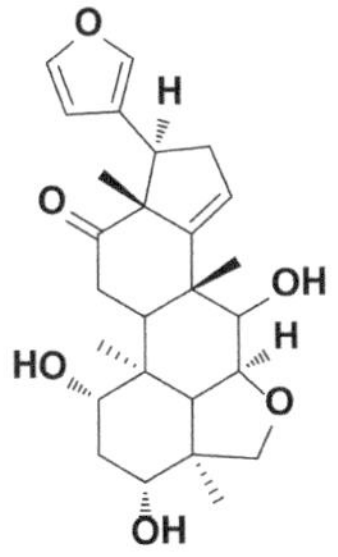

NIMBIDININ

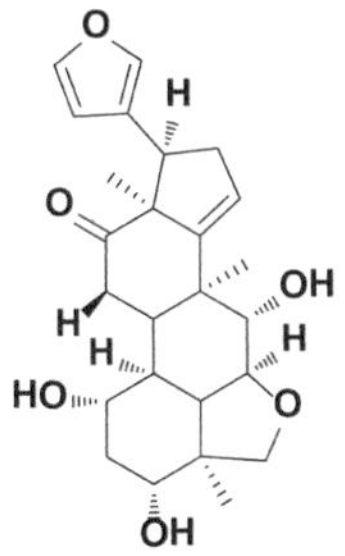

VILASININ

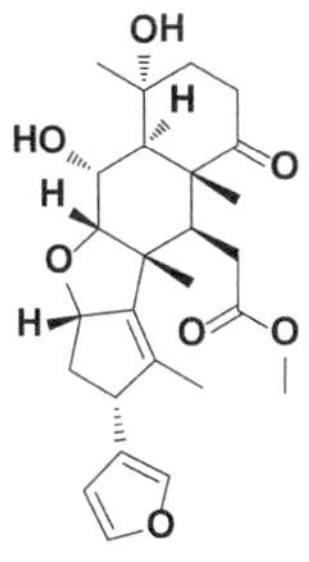

NIMBANDIOL

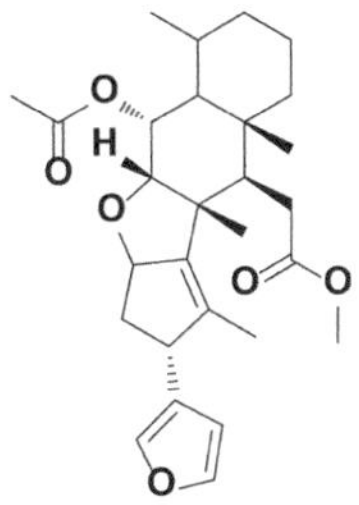

NIMBINENE

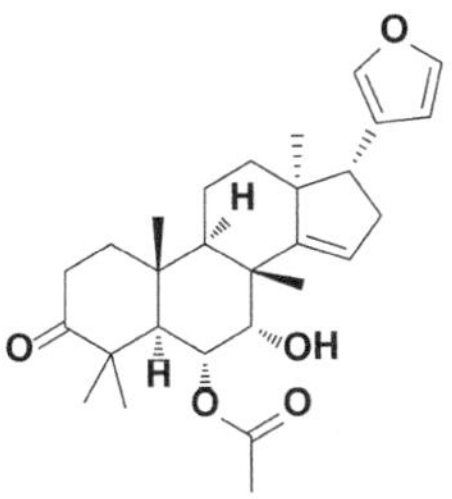

MELDENIN

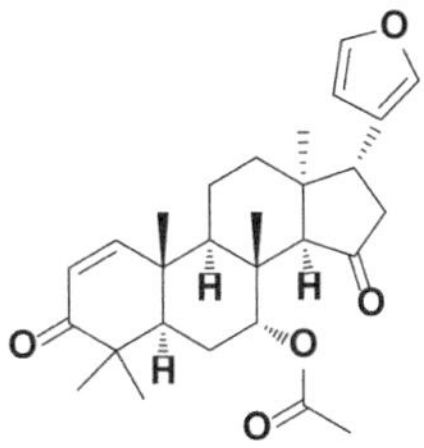

7-ACETYL NEOTRICHILENONE

FIGURE 1.1   *(Continued).*

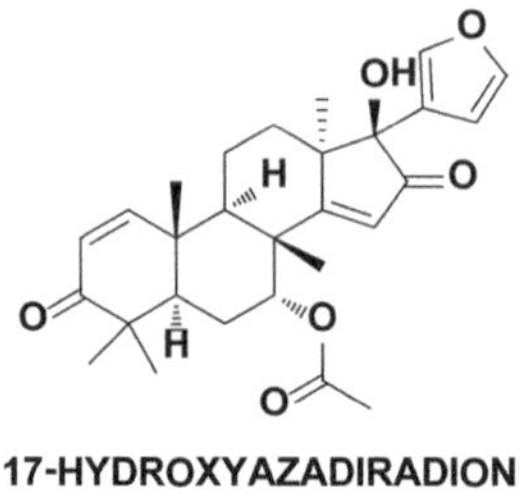

17-HYDROXYAZADIRADIONE

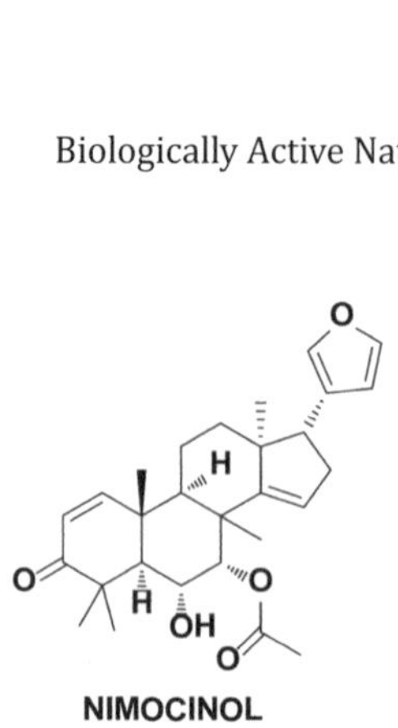

NIMOCINOL

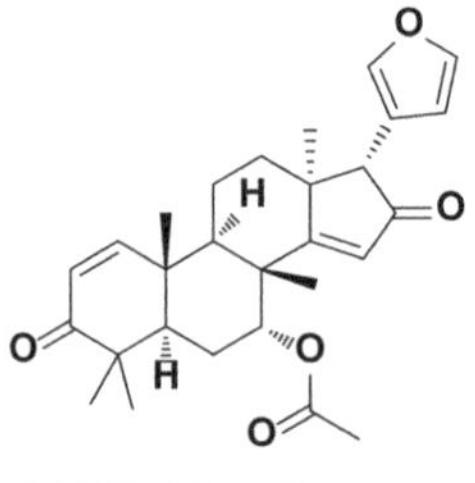

AZADIRADIONE

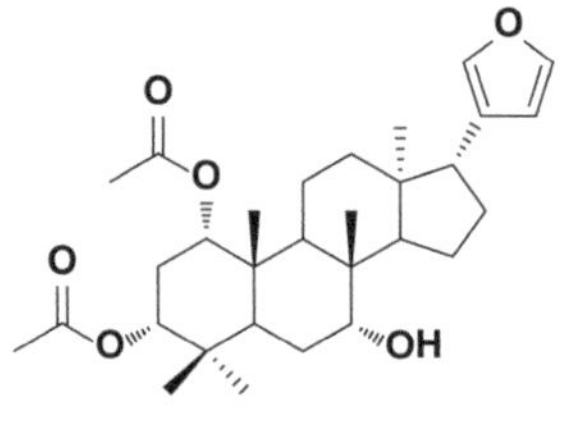

1,3-DIACETYL VILASNIN

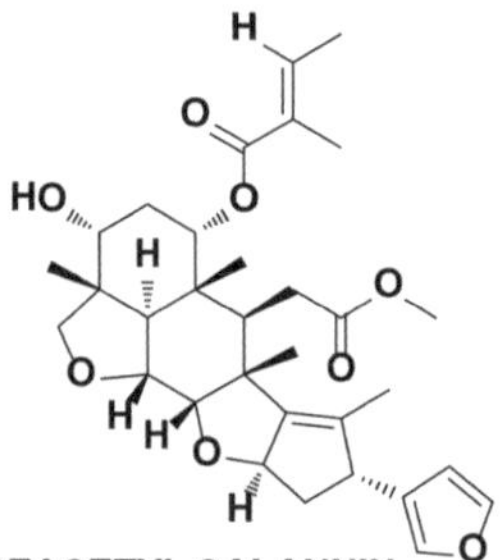

DEACETYL SALANNIN

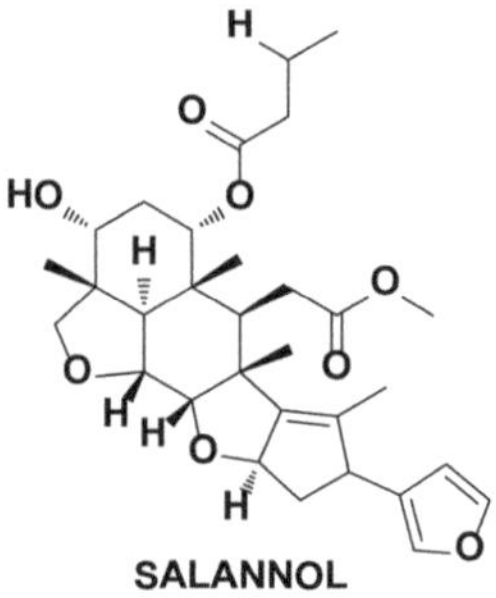

SALANNOL

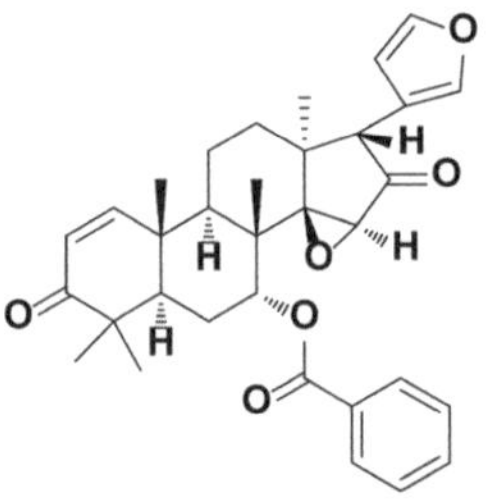

7-DEACETYL 7-BENZOYLEPOXYAZADIRADIONE 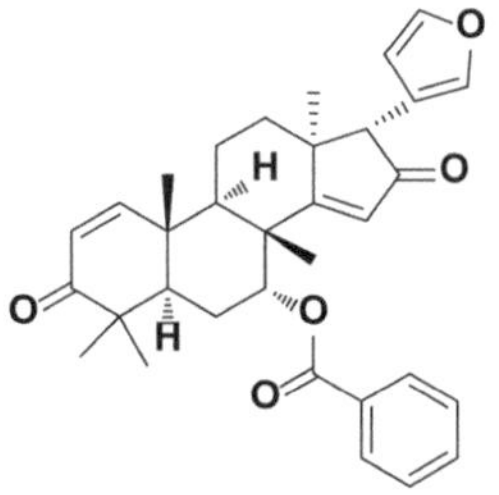   7-DEACETYL 7-BENZOYLAZADIRADIONE

FIGURE 1.1   *(Continued).*

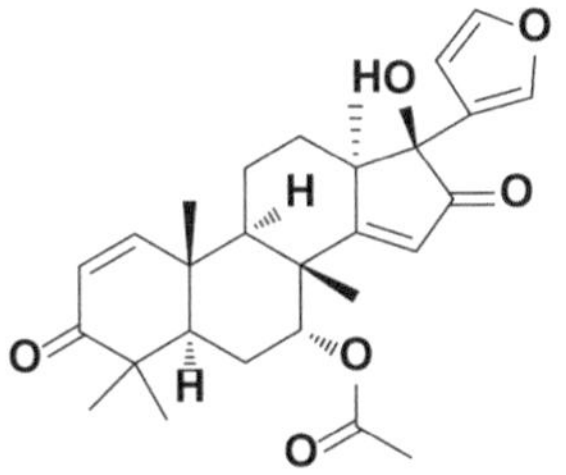

17BETA-HYDROXYAZADIRADIONE

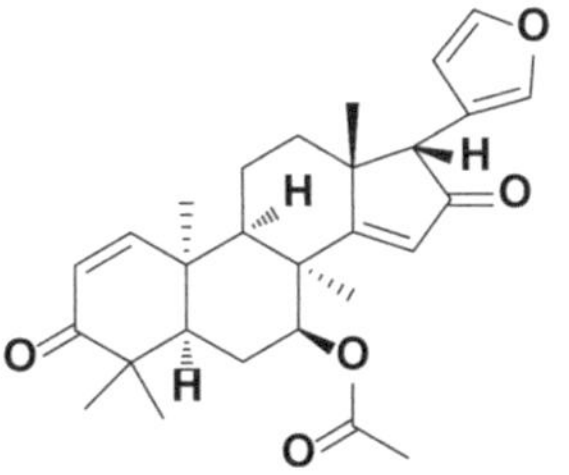

17EPI-HYDROXYAZADIRADIONE

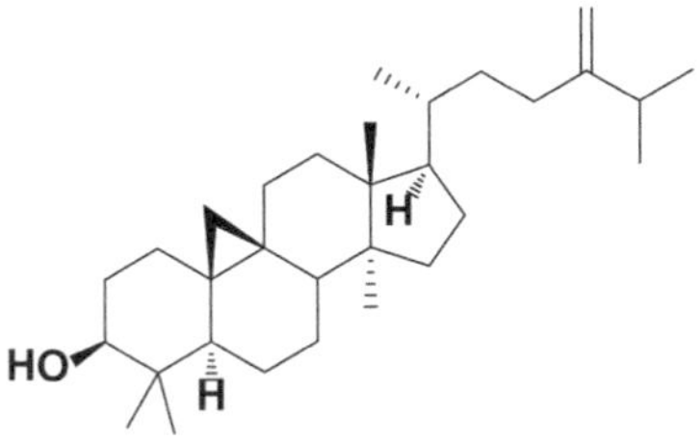

24-METHYLENE CYCLOARTANOL

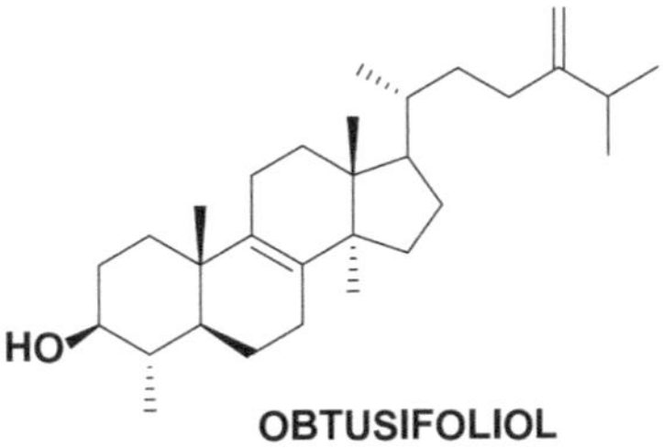

OBTUSIFOLIOL

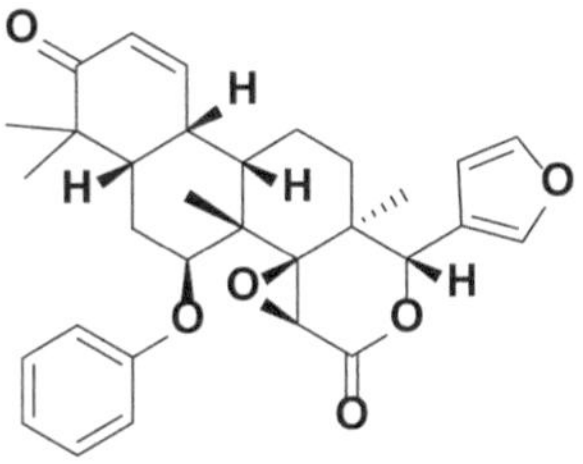

7-DEACETYL 7-BENZOYLGEDUNIN

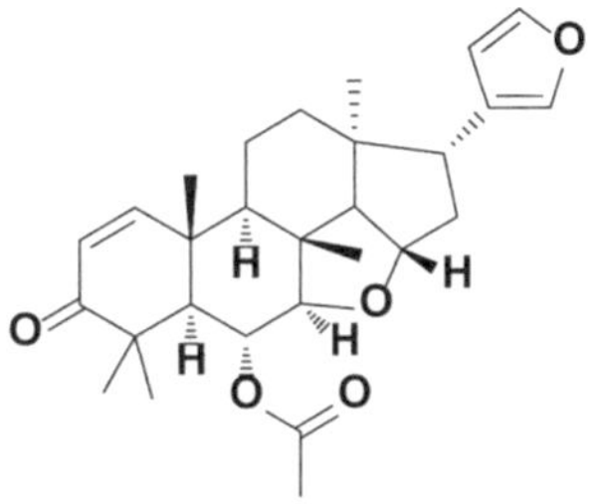

VEPININ

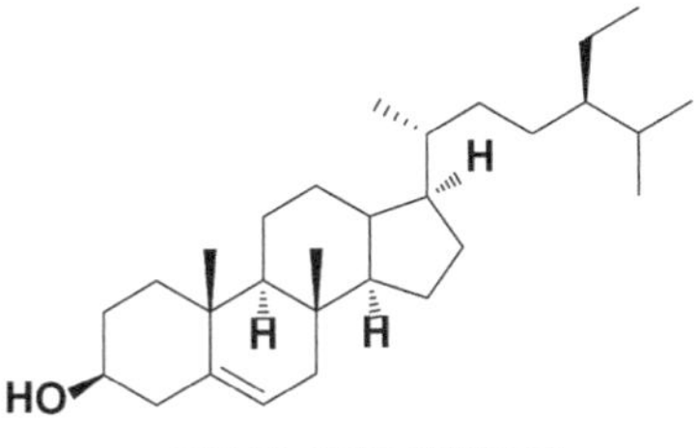

BETA-SITOSTEROL

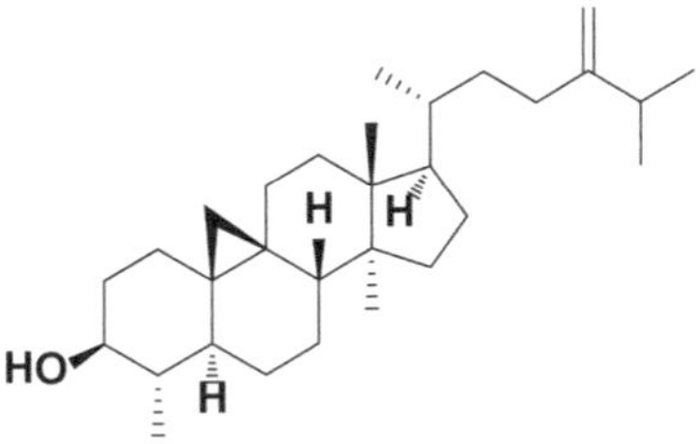

CYCLOEUCALENOL

FIGURE 1.1   *(Continued).*

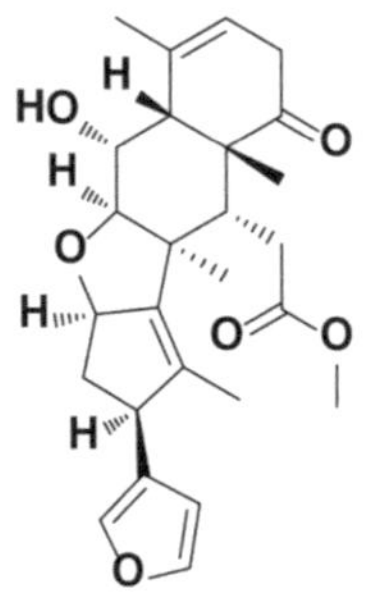

6-DEACETYL NIMBINENE

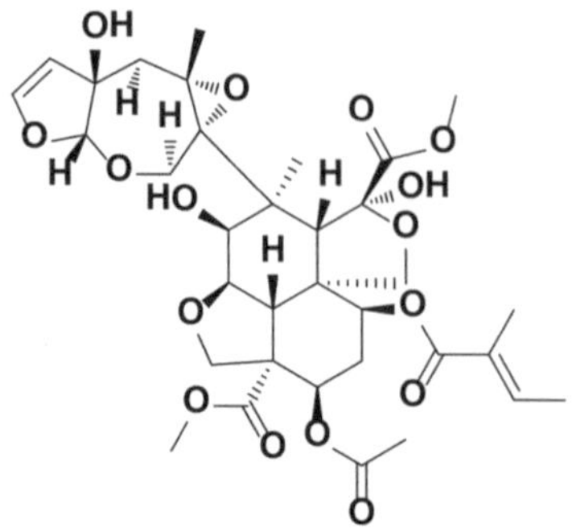

AZADIRACHTIN

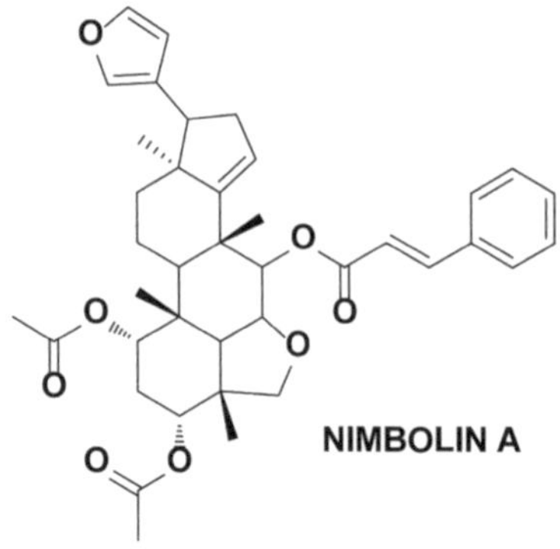

NIMBOLIN A

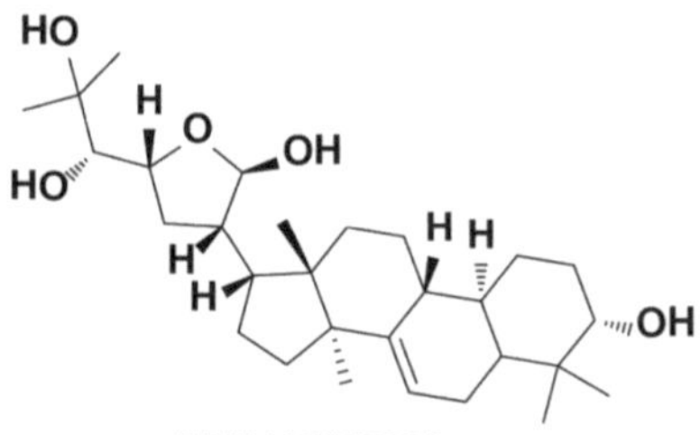

MELIANTRIOL

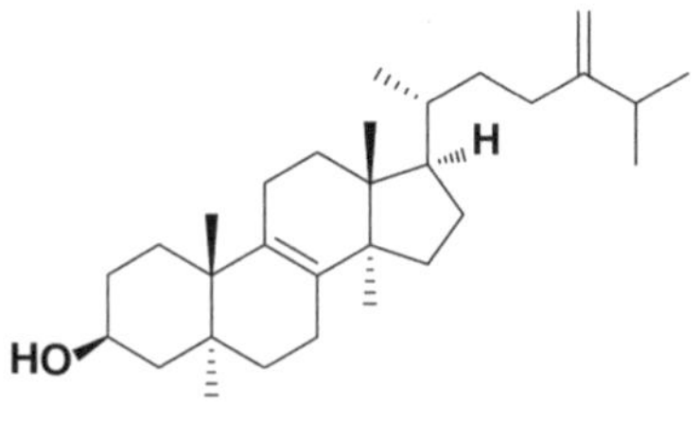

4alpha-methyl-5alpha-ergosta-8,24(28)-dien-3ß-ol

FIGURE 1.1   *(Continued).*

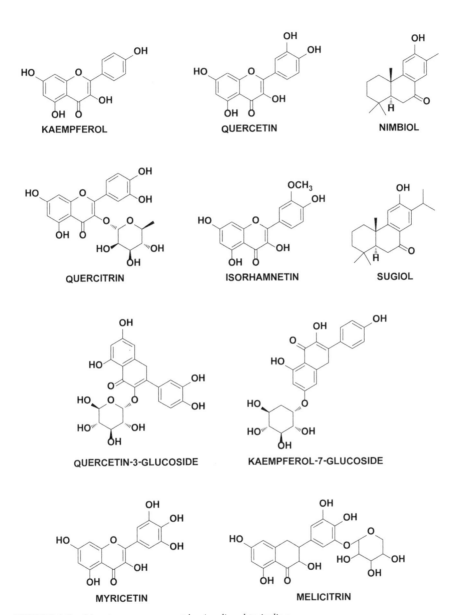

FIGURE 1.2   Non-terpenes present in *Azadirachta indica*.

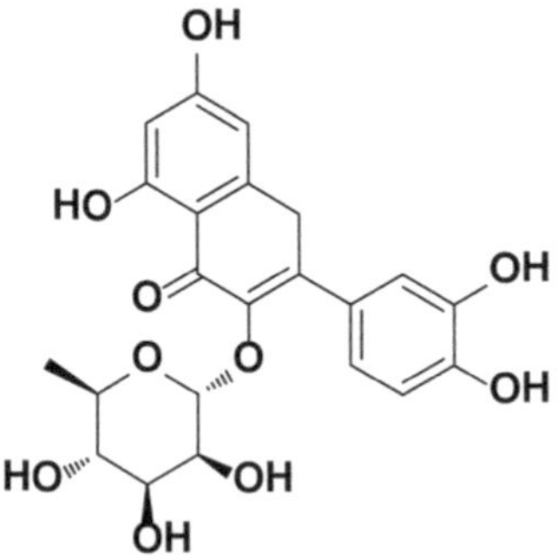

QUERCETIN-3-O-RHAMNOSIDE

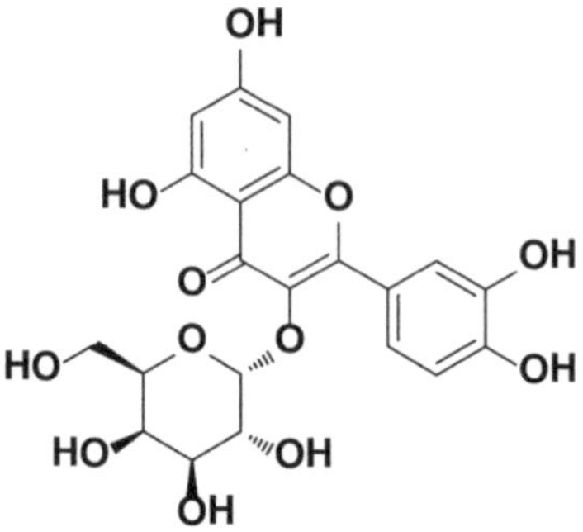

HYPERIN

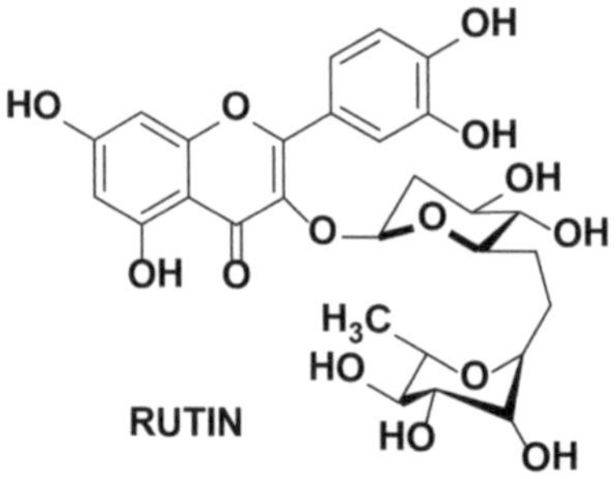

RUTIN

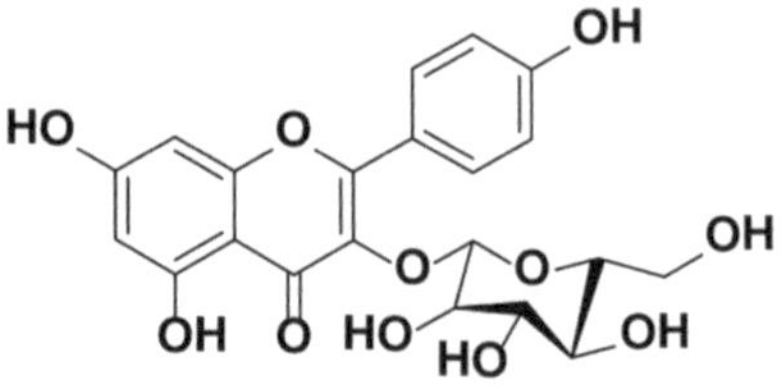

ASTRAGALIN

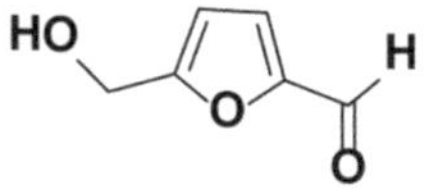

5-HYDROXY METHYL FURFURAL

FIGURE 1.2   *(Continued).*

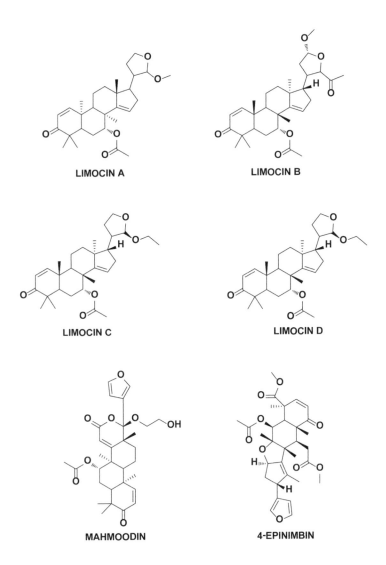

FIGURE 1.3 Miscellaneous compounds present in *Azadirachta indica*.

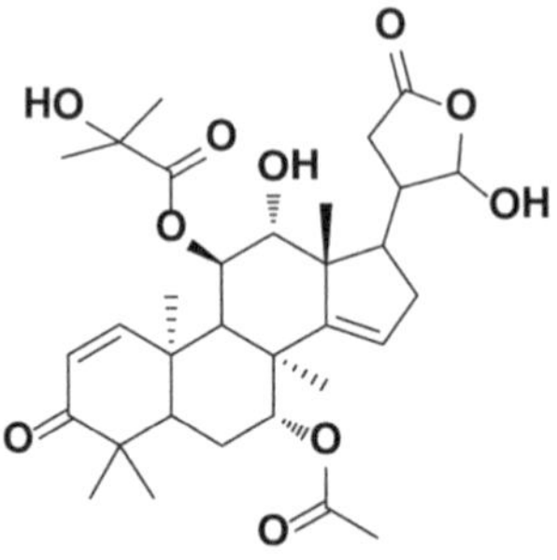

ISONIMBOCINOLIDE

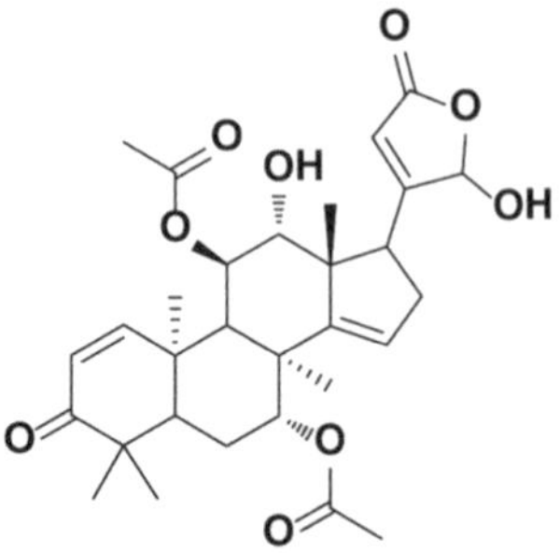

ISONIMBOLIDE

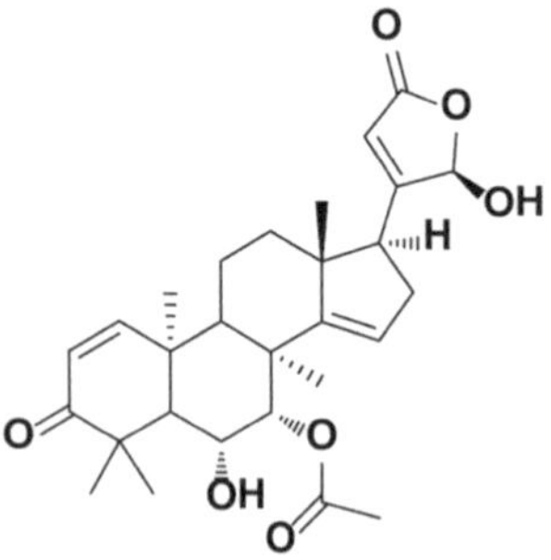

ISONIMOCINOLIDE

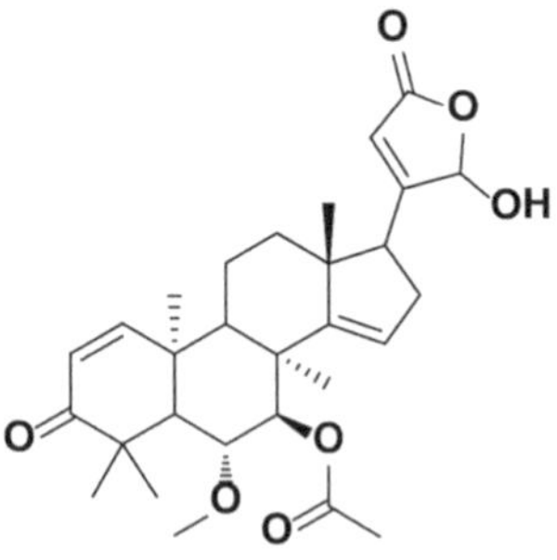

ISONIMOLIDE

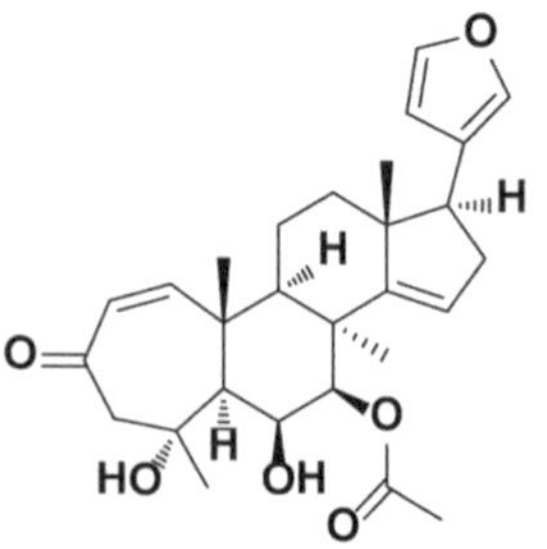

4ALPHA-6ALPHA-DIHYDROXY-
ALPHA-HOMOAZADIRONE

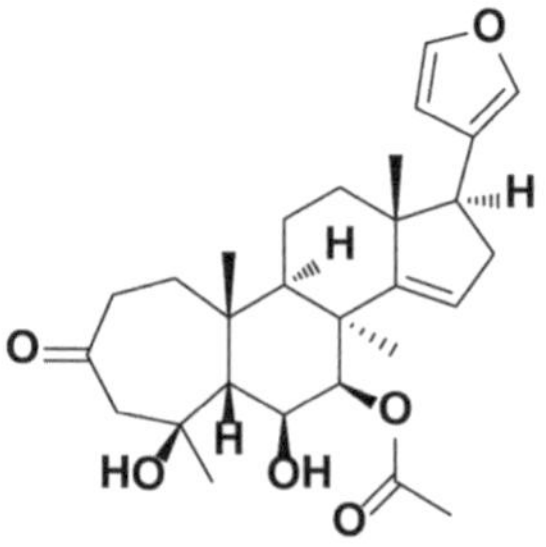

4ALPHA-HYDROXY-ALPHA-
HOMO-ISOMELDENIN

FIGURE 1.3   *(Continued).*

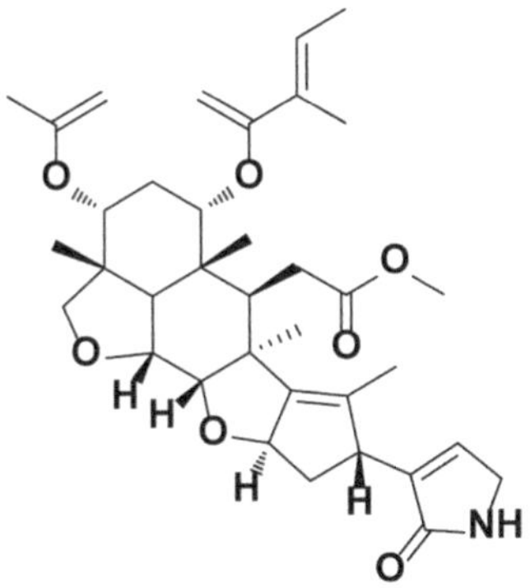

SALANOLACTAME-I

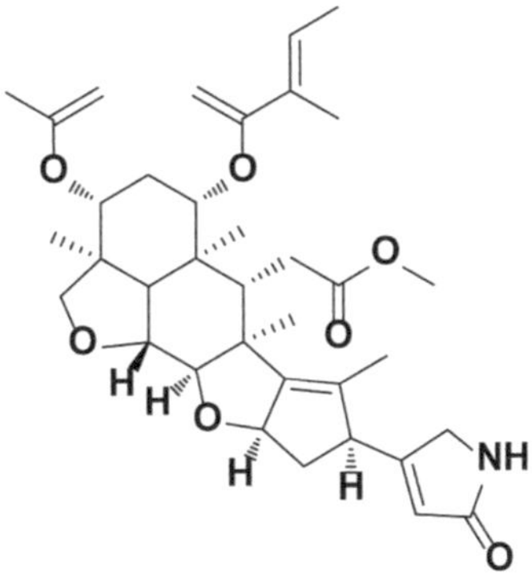

SALANOLACTAME-II

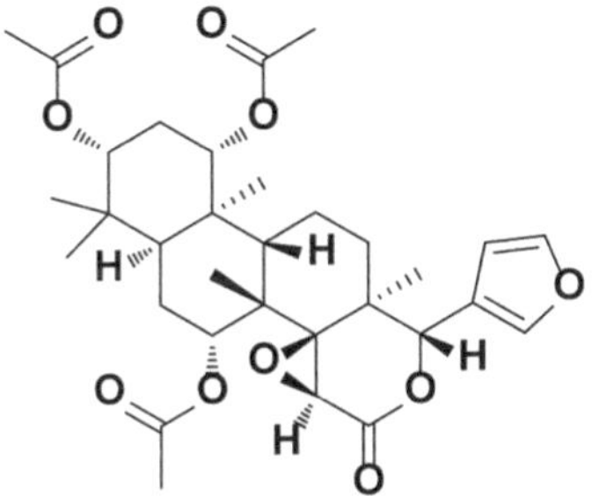

KHIVORIN

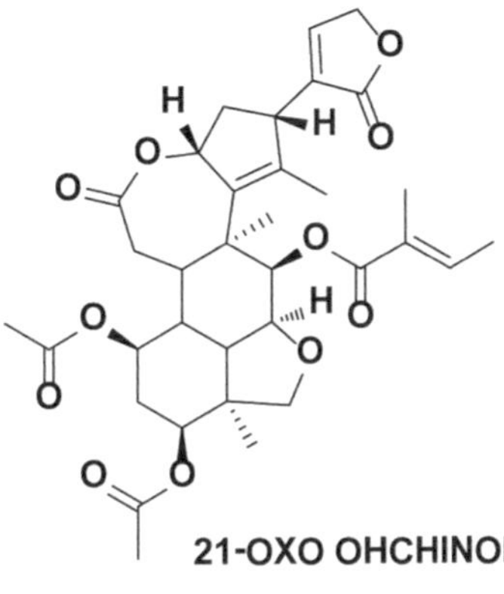

21-OXO OHCHINOLIDE

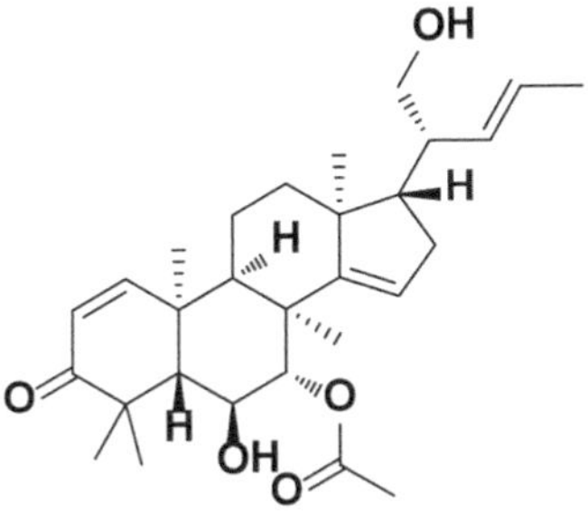

ZEESHANOL

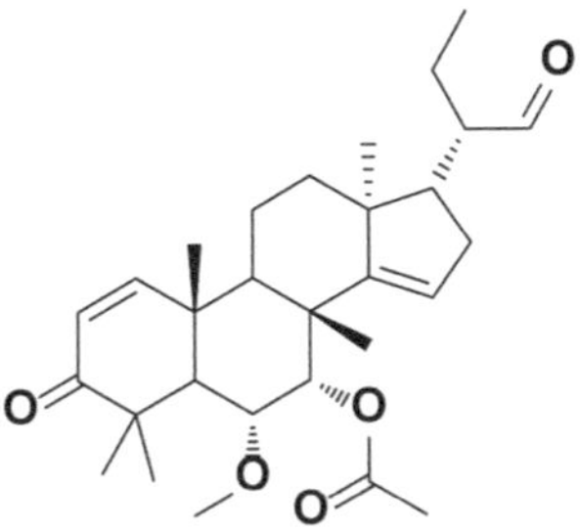

ZAFARAL

FIGURE 1.3 *(Continued).*

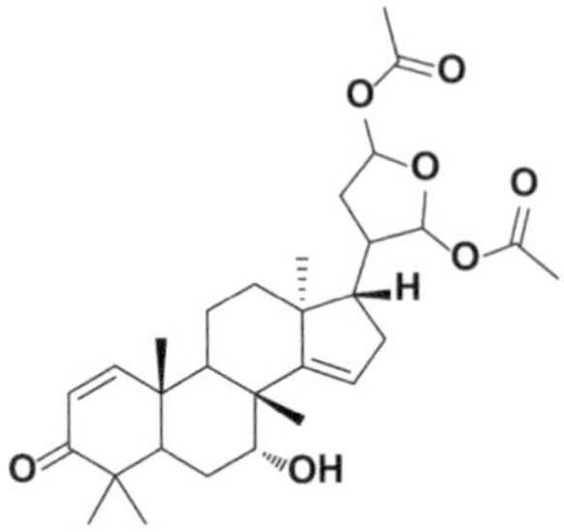

SALIMUZZALIN

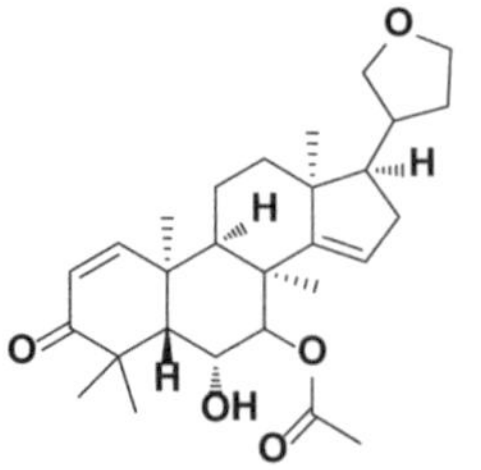

22,23-DIHYDRONIMOCINOL

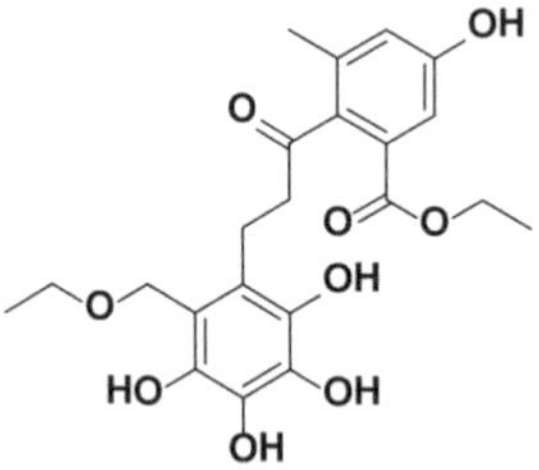

NIMBOCHALCIN

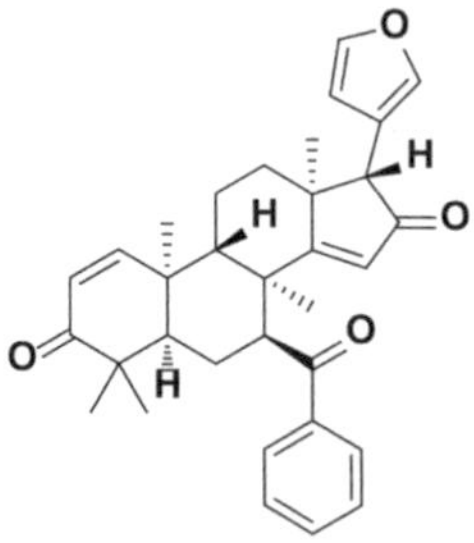

AZADIRADIONEBENZOATE

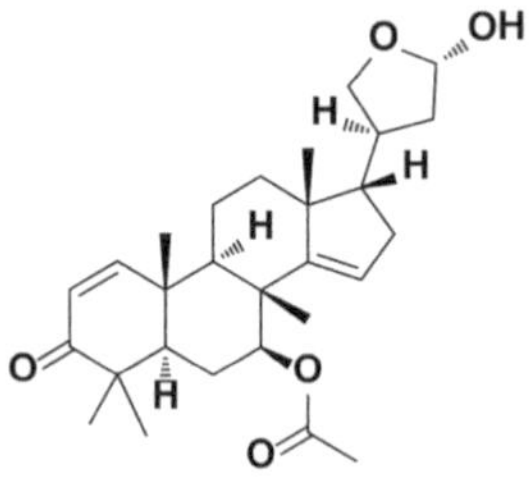

23-DESMETHYLLIMOCIN

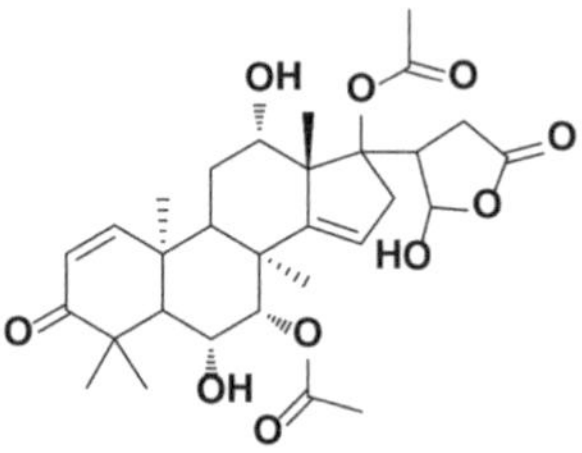

ISONIMOLICINOLIDE

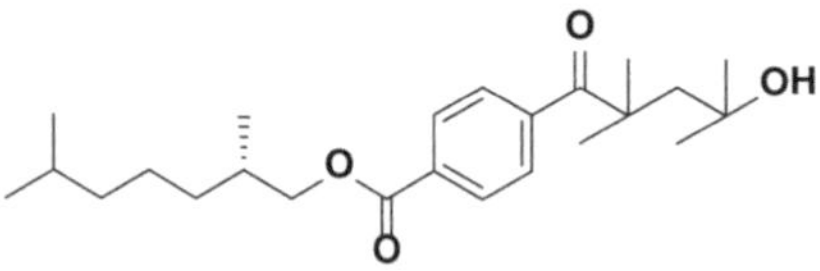

NIMBOCETIN

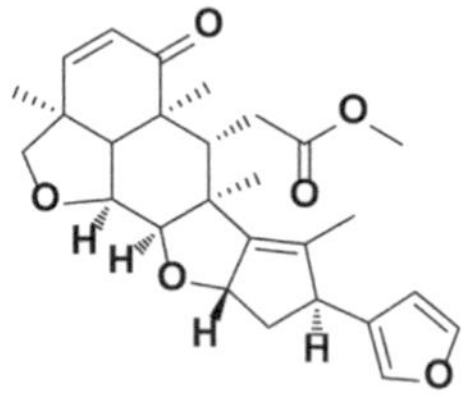

28-DEOXONIMBOLIDE

FIGURE 1.3   *(Continued).*

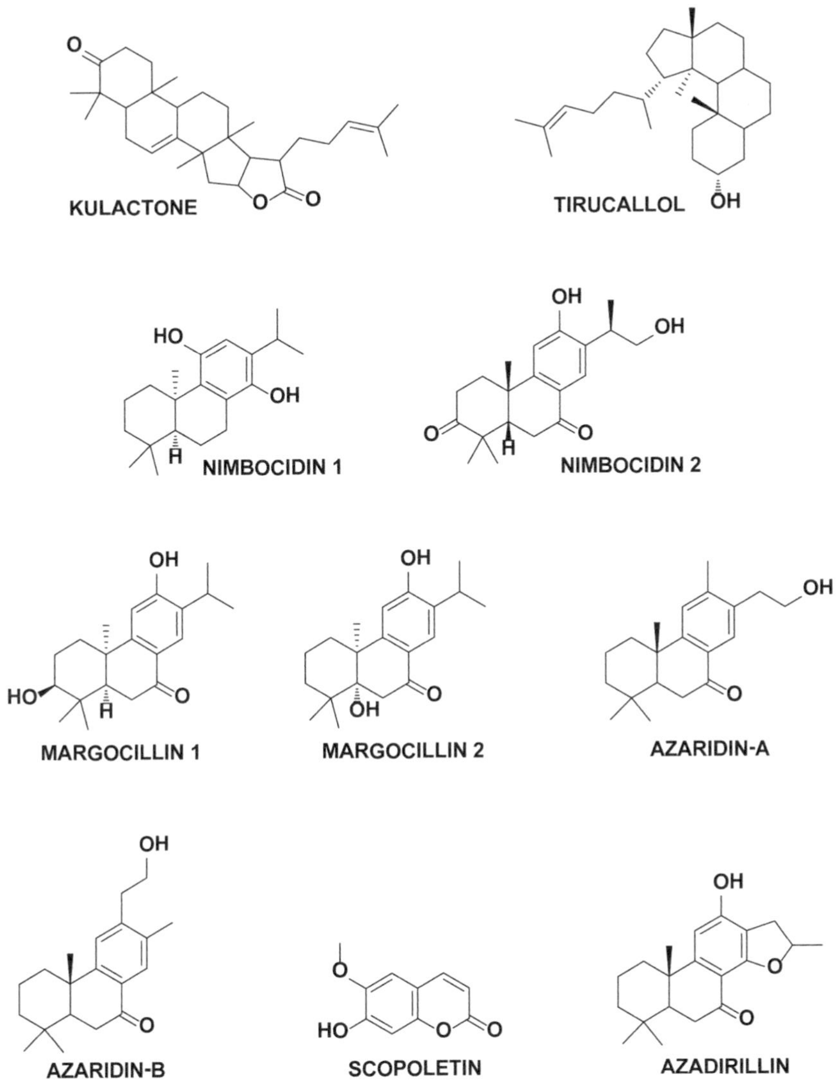

FIGURE 1.3 *(Continued).*

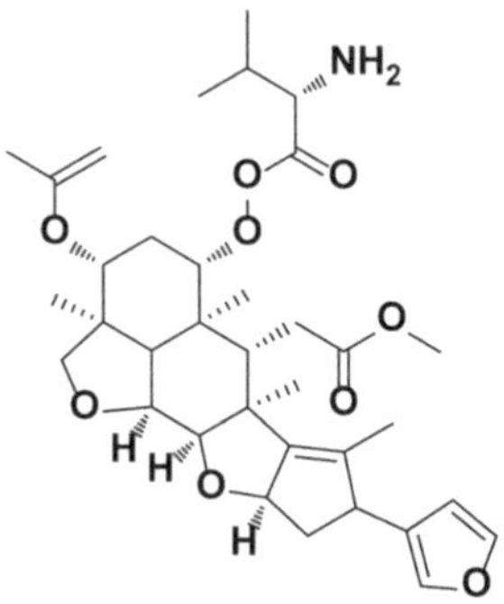

SALANNOLACETATE

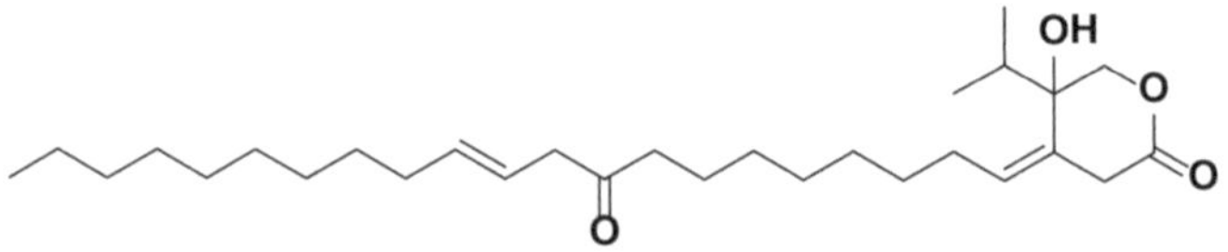

MARGOSINONE

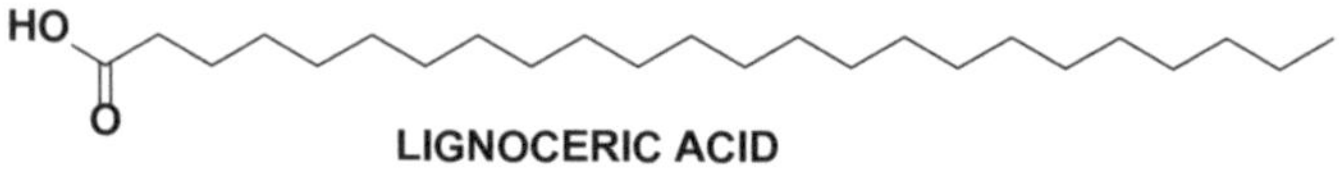

LIGNOCERIC ACID

FIGURE 1.3   *(Continued).*

1.3   PHARMACOLOGICAL ACTIONS OF NEEM

In Ayurveda, neem is used extensively. All the parts of the Neem tree are used as a traditional medicine (TM) in the form of home remedies against commonly occurring human ailment. Table 1.1 lists the pharmacological activities exhibited by the different parts of the neem tree. This tree is used in a massive amount and this is because of its availability throughout the year.

1.3.1   *HEART DISEASE*

Neem detains coagulation of blood, relaxes erratic heartbeats, and helps in decreasing the elevation of heart rate and increased blood pressure [17].

TABLE 1.1   Medicinal Properties of Different Parts of Neem

Parts	Biological Activity
Leaf	• Antifungal
	• Antibacterial
	• Antimalarial
	• Anti-inflammatory
	• Analgesic
	• Anticancer
	• Antipyretic
	• Anti-infertility
	• Antigenotoxic
	• Hepatoprotective
	• Immunostimulant
	• Orodental protection
Bark	• Antibacterial
	• Antimalarial
	• Anti-inflammatory
	• Antiulcer
	• Anti-stimulant
	• Anticancer
	• Hepatoprotective
Flower	• Antioxidant
	• Anticancer
Seeds	• Antimalarial
	• Antioxidant
	• Antifertility
	• Anticancer
Oil	• Antifungal
	• Antifertility
	• Antipyretic
	• Anti-hyperglycemic

1.3.2   SKIN DISEASES

Neem has got a significant effect on the skin conditions that have reached a chronic stage. Conditions including psoriasis and acne; ringworm, eczema, even a stubborn skin disease like warts can be treated with a high quality of neem oil. In Siddha medicine, treatment of skin disease is done by both neem leaves and oil. It is an excellent component to rejuvenate the skin and can be classified as the best natural cosmetic in the form of Neem oil.

1.3.3   INSULIN SENSITIZATION

Extricates of Neem that are taken orally which decreases the insulin requirements by 30–50% in insulin-sensitive diabetic and non-ketonic patients.

1.3.4   AIDS

The National Institute of Health (NIH), USA stated that the extracts from neem kill AIDS. Viruses and patients have been handed awards for these extracts as a treatment for AIDS [18].

1.3.5   PERIODONTAL DISEASE

Researchers from German have proven that the extracts from neem prevent periodontal diseases and tooth decay which leads to good oral health [19]. These extracts from Neem leaf have an antimicrobial effect on *Enterococcus faecalis* and *Candida albican*. Therefore, it can be a potential irrigant of endodontic related ailments [20].

1.3.6   SEXUALLY TRANSMITTED DISEASES

Few researchers have highlighted the efficacy of Neem in the treatment of sexually transmitted diseases and the responses have been overwhelming and positive. In the case of *Neisseria gonorrhoeae* infection, the efficacy of Neem is proven to be best for treatment.

1.3.7   DUODENAL LESIONS

Neem extracts give prominent protection from discomfort, give fast relief, speedy duodenal lesions, and quickly heals infections [21].

1.3.8   GYNECOLOGICAL PROBLEMS

Hot water extricates of the neem's bark is consumed orally by the adult females in the forms of emmenagogue and tonic.

1.3.9   LEPROSY

Leprosy can be treated by a regular dosage of neem, Anthraquinone fraction of dried fruit, flower, and leaves can be consumed orally. Hot water extracts of the flowers and leaves of this tree are taken orally because they act like an anti-hysteric remedy and also used for treating external wounds [22].

1.3.10   DIABETES

To cure diabetes, dried flower is taken orally. Hot water extricates of neem's dried fruit is used for treating patients with piles and for external ailments like ulcers and skin diseases. Hot water extracts of all the parts of this plant are used as an anthelmintic, an insecticide purgative and as an anthelmintic. Juices taken from the barks of *Andrographic paniculata, A. indica,* and *Tinospora cardifolia* are consumed orally for treating filariasis. The hot water extract is also taken for commonly prevailing diseases like fever, diabetes, in the forms of tonics, refrigerants, and anthelmintics [23].

1.3.11   COMMON FEVER

A combination of fruit leaf and roots are mixed with dried ginger or "Triphala" is taken orally with warm water to treat common fever. Leaves of this tree are preserved with woolen and other types of clothes to prevent their insecticidal properties. Juice obtained from Neem leaves are given in gonorrhea and leucorrhea. Also, neem leaves are applied externally as a poultice to relieve and treat boils, their infusions are used as an antiseptic wash to promote the healing of wounds and ulcers [24].

1.3.12   SKIN RELATED AILMENTS

A paste-form of Neem leaves is used to manage dermatological ailments like external wounds and fungal infections like ringworms and eczema. It is considered that taking a bath from Neem leaves is beneficial for curing commonly prevailing problems like Itching and other skin diseases. Juice of Neem leaves are used as nasal drops to treat worm infestation found in the nose. Steam Inhalation of Neem's bark is useful to cure inflammation occurring commonly in the throat. Decoctions of Neem can be used for treating intermittent fever, general debility convalescent, and loss of appetite after fever. An infusion made out of neem flowers is given in dyspepsia and in general debility [25]. The fragile twigs of the neem tree are used as a toothbrush, which is believed to maintain the systems of the body and keep it healthy. Also, it is used for keeping the breath fresh, mouth clean, and eliminate any false sweat smell. The seed oil of neem is used in diseases like leprosy, syphilis, eczema, and chronic ulcers.

1.3.13   LARVICIDAL ACTIVITY

Testing was done on four plants by taking their aqueous extracts for characterizing their larvicidal properties. Initially, laboratory-reared larvae were kept open to 1 to 5 ppm concentrations of the extricates of *A. indica, Gymnema sylvestre, Nerium indicum,* and *Datura metel* which showed that the *A. indica* elicited 70–99% mortality, followed by *G. sylvestre* 44–89%, *N. indicum* 41–74%, and *D. metel* elicited 19–54% mortality to larvae. The extracts of *A. indica* and *G. sylvestre* were found to be significantly effective in controlling *Culex* larvae[26]. The formulation of neem oil was identified to be effective in controlling mosquito larvae in various breeding sites under certain natural field conditions. Neem oil formulations are relatively less toxic, more eco-friendly. Insects were unable to develop resistance and may be used as an alternative to other pesticides for control of vector-borne diseases [27].

1.3.14   ANTI-VIRAL ACTIVITY

Extracts of aqueous Neem leaf exerts an antiviral activity opposite to the *Vaccinia* virus, *Chikungunya*, and *Measles* virus (*in vitro*) [38]. Rao et al. [39] found that a 10% water extract of tender leaves, exhibits antiviral activity against vaccinia and variola viruses. Virucidal and antiviral effects of the methanolic extract taken from Neem leaves (NCL-11) have been demonstrated against group-B-coxsackie viruses [40]. Parida et al. [41] reported the stoppage of the potential of Neem leaves on Dengue virus type-2 replication.

Aqueous neem leaf extricates showed low to moderate suppression of the viral DNA polymerase of the hepatitis B virus [42].

1.3.15 ANTI-BACTERIAL ACTIVITY

Methanol, petroleum ether, and aqueous extracts of the leaves of *A. indica* (Meliaceae), bulbs of *Allium cepa* (Liliaceae), and methanol extract of gel of *Aloe vera* (Liliaceae) were all together screened for their anti-microbial activity respectively, using the method of cup plate agar diffusion. They all were tested against six bacteria; four Gram-negative bacteria (*Escherichia coli, Proteus vulgaris, Pseudomonas aeruginosa*, and *Salmonella typhi*) and two Gram-positive bacteria (*Bacillus subtilis* and *Staphylococcus aureus*). The sensitivity of the microorganisms to the extricates of these plants were compared with each other and with a handful of antibiotics. The methanol extract of *A. indica* was an illustration of pronounced activity against *B. subtilis* (28 mm) [28].

1.3.16 HYPOGLYCEMIC AND HYPOLIPIDEMIC POTENTIALS

The hypoglycemic action of *A. indica* was been examined in diabetic rats. After treating these rats for 24 hrs, *A. indica* 250 mg/kg (single-dose study) reduced glucose by 18%, cholesterol by 15%, triglycerides by 32%, urea by 13%, creatinine by 23%, and lipids by 15%. A Multiple dose study was performed for 15 days which showed reduced levels of urea, lipids, triglycerides, creatinine, and glucose. In a glucose tolerance test, diabetic rats were treated with neem extricates 250 mg/kg demonstrated that glucose levels were comparatively less when compared to the control group. *A. indica* prominently reduced glucose levels on the 15th day in the rats with diabetes [28].

1.3.17 ANTI-OXIDANT

Extricates from young leaves and flowers hold strong and higher antioxidant potential in them. An indicator of oxidative stress, malondialdehyde (MDA) was reduced by 46.0% and 50.6% for flower and leaf-based extracts, respectively, prompting the recommendation to use neem as a vegetable bitter tonic to promote good health [29].

1.3.18 ANTI-ULCER

Extracts from the Neem tree's bark reduces human gastric acid hypersecretion, gastro-esophageal, and gastroduodenal ulcers. Also, after 10 weeks, the duodenal ulcers nearly get completely healed, after 6 weeks, one case of esophageal ulcer and gastric ulcer were noted to be completely healed [30].

1.3.19 ANTI-MALARIAL ACTIVITY

The antimalarial activities of the tablet suspension, containing bark and leaf of *A. indica* were evaluated on *Plasmodium yoellinigeriensis* on infected mice. After treatment, the tablet suspensions exhibited high prophylactic, moderate suppression, and a very minimum curative schizonticidal effect. The tablet suspensions from the leaf and bark at a concentration of 800 mg/kg and chloroquine at a concentration of 62.5 mg/kg body weight produced an average percentage of parasitemia, i.e., 79.6%, 68.2%, and 99.5% for leaf, bark, and chloroquine, respectively (in chemo-suppression). Also, in the prophylactic treatment, the tablet suspensions at 800 mg/kg and pyrimethamine at a concentration of 0.35 mg/kg gave an average parasitemia reduction of 75.3%, 65.6%, and 98.3% for the leaf, bark, and pyrimethamine. There was a clear indication of results to give a moderate beneficial effect [31].

1.3.20 ANTI-FERTILITY EFFECT

Neem and seed extracts consumed orally at the beginning of the post-implantation stage was observed to result in pregnancy termination in rodents and primates, this had no permanent effects. The mechanism of action was believed to be not fully understood [32]. Praneem (licensed to Panacea Biotec Ltd., India) is a poly-herbal vaginal tablet that has been studied and proved to be very effective in immobilizing sperms [33, 34].

1.3.21 ANTI-DENTAL CARIES

It has been tested and proved that a Neem-extract dental gel, very prominently reduces plaque and bacteria (*Streptococcus mutans* and *Lactobacilli* species were tested) commercially available mouthwash containing the germicide chlorhexidine gluconate (0.2% w/v) [35]. In preliminary findings, Neem inhibited *S. mutans* (bacterium causing tooth decay) and reversed incipient carious lesions that are primary dental caries.

1.3.22   ANTI-HYPERTENSIVE

A study was done and further resulted showing that the administration of aqueous extract of Neem with DOCA salt shows the prevention of developing hypertension in rats. Administration of the mature leaf extract decreased serum cholesterol significantly without changing serum protein, protein urea, and uric acid level in rats [36, 37].

1.3.23   ANTI-CARCINOGENIC

The term long known by the Asians as 'nasoor' in the olden texts is referred to as the "tumor." These tumors are observed to be cured by the extracts of neem. Cancer cells are characterized by multiple attributes, which includes excessive cell growth, reactivation of energy metabolism that holds up the unpowered proliferation, resistance to cell death, immortality, induction of angiogenesis, the capacity to occupy and metastasize to far points and repression of the immune response against tumor cells as proven by many studies, preclinical studies have manifested gathering evidences implicating that the anticancer effects of neem are communicated through modulation of many cellular processes [43].

Neem has the potential ability to withdraw the cancerous phenotype (tumor). In the last two decades, across the globe especially in India, scientists have gathered data, which explained the primary cause of cancerous phenotype due to mutagens and pro-carcinogens, which now are believed to be treated by the extricated units of the neem tree. Dietary doses of aqueous neem leaf during the chemopreventive therapy were studied on the murine system that was *in vivo* against ^{3}H-B-α-P (Benz-α-pyruvate)-prompt initiation of cancer that was measured in terms of ^{3}H-B-α-P-DNA adduct. The results obtained after the study implied that the extracts of Neem leaf reduced the metabolic start of ^{3}H-B-α-P with a consequential reduction in the level of ^{3}H-B-α-P-DNA adduct development. These biochemical and molecular modulators seen at the very start of carcinogenesis, highlight the chemopreventive importance of *A. indica* extracts.

Active components in the form of many chemical compounds found in. leaves, barks, seeds, and seed oil decreases cancers and tumors very efficiently with producing absolutely no side effects. Compounds which are biologically very active (polysaccharides, terpenoids, steroids, and limonoids) have been profusely used for treating many types of cancer conditions. Different components of Neem restrict proliferation, Initiates apoptosis, other types of cell death, and decrease oxidative cellular stress. The expression of genes carrying out many cellular processes is adjusted to be altered in response to produce extracts of neem in carcinogen-induced hamster buccal pouch (HBP) model. Tumor in the microenvironment plays a highly important role in angiogenesis and metastasis. Tumor cells show the potential to regulate their local environment (or microenvironment), this triggers inflammation, induces angiogenesis, and supports cell invasion. Therefore, tumor microenvironment frisks significant roles in the initiation and further progression of tumors. Engrossingly, components of neem seem to modulate tumor microenvironment through multiple mechanisms including attenuation of angiogenesis and improved cytotoxicity.

There is significant evidence that is interesting and compelling to suggest that neem can be used as a tumor suppressor. On research in India, Europe, and Japan, this has been found that limonoids and polysaccharides present in the bark, leaves, and seed oil of neem decreased cancers and tumors and correspondently showed its efficacy against lymphocytic leukemia. Presently, In Japan, hot water components from neem bark showed significant improvement against several types of tumors. Few of these extracts were similar to or better than the normal anticancer agents, specifically against tumors which are solid [44].

Programmed cell death or apoptosis is a genetically potentiated process that has been kept in conserved in bulk among the metazoans from the past several decades. It is considered to be the finest natural procedure of surgery that requires blades that biochemically processed and led by a host of caspase enzymes in order to wipe out the irreparable, wayward, incorrigible, redundant, and unwanted cells of the body which is done without shedding even a single drop of blood or any material loss to the body all in all. The modern medical practitioners have obtained the elimination of cancer cells via orchestration of apoptosis as a right therapy of choice [45].

1.3.24 DRUG METABOLIZING ENZYME

The preparations of Neem in some cases have been studied to activate antitumor activities of some drugs by offering protection against life warning side-effects of the agents which are chemotherapeutic. Pre-treatment of Swiss mice with Neem extract not only reduced neutropenia and leucopenia but also enhances the antitumor activities of cyclophosphamide consequently [45].

1.4   CONCLUSION

Neem is one of its unique types of plant which has versatile pharmacological attributes and is correctly named as "The Doctor Tree." This plant is one of the most promising plants of all times with a fact that basically, it can be of great help to each living human body on the Earth, In one way or other, this tree is an ultimate blessing on the face of the earth.

KEYWORDS

- *Azadirachta indica*
- ethnobotany
- ethnopharmacological
- neem
- pharmacology
- phytochemistry

REFERENCES

1. Abdel-Ghaffar, F., Al-Quraishy, S., Al-Rasheid, K. A. S., & Mehlhorn, H., (2012). Efficacy of a single treatment of head lice with neem seed extract: An *in vivo* and *in vitro* study on nits and motile stages. *Parasitol. Res., 110*, 277–280.
2. Girish, K., & Shankara, B. S., (2008). Neem-a green treasure. *Electron J. Biol., 4*(3), 102–111.
3. Singh, K. P., (1997). Medicinal properties of mulberry : A review. *Indian Drugs, 34*(9), 488–492.
4. Sateesh, M. K., (1998). Microbiological investigations on die-back disease of neem (*Azadirachta indica* A. Juss.). *PhD Thesis*. University of Mysore. Mysore, India.
5. Thakkar, P. S., (1997). Editorial notes. *Global Neem Update, 2*, 1.
6. Parotta, J. A., (2001). *Healing Plants of Peninsular India* (pp. 495–496). New York, CABI Publishing.
7. Brahmachari, G., (2004). Neem-an omnipotent plant: A retrospection. *Chem. Biochem., 5*, 408–421.

8. Kumar, R. V., & Gupta, V. K., (2002). Thrust on neem is need of today. In: *Employment News*. New Delhi, India.

9. Chari, M. S., (1996). Neem and transfer of technology. In: Singh, R. P., Chari, M. S., Raheja, K., et al., (eds.), *Neem and Environment* (Vol. I). Oxford and IBH Publishing Co. Pvt. Ltd., New Delhi, India.

10. Jattan, S. S., Shashikumar, Pujar, G., et al., (1995). Perspectives in intensive management of neem plantations. *Indian For., 121*, 981–988.

11. Tewari, D. N., (1992). *Monograph of Neem (Azadirachta indica A. Juss.).* International Book Distributors, Dehra Dun, India.

12. Hegde, N. G., (1995). Neem and small farmers-constraints at grass root level. *Indian For., 121*, 1040–1048.

13. Kirtikar, K. R., & Basu, B. D., (1975). In: Blatter, E., Cains, J. F., & Mhaskar, K. S., (eds.), *Medicinal Plants* (p. 536). Vivek Vihar, New Delhi.

14. Maithani, A., Parcha, V., Pant, G., Dhulia, I., & Kumar, D., (2011). *Azadirachta indica* (neem) leaf: A review. *J. Pharm. Res., 4*(6), 1824–1827.

15. Hashmat, I., Azad, H., & Ahmed, A., (2012*). Neem (Azadirachta indica A. Juss): A Nature's Drugstore: An Overview, 1*(6), 76–79.

16. Atawodi, S. E., & Atawodi, J. C., (2009). *Azadirachta indica* (neem): A plant of multiple biological and pharmacological activities. *Phytochem. Rev., 8*(3), 601–620.

17. Bandyopadhyay, U., Biswes, K., Sengupta, A., Moitra, P., Dutta, P., et al., (2004). Clinicak studies on the effect of neem (*Azadirachta indica*) bark extract on gastric secretion and gastro duodenal ulcer. *Life Sci., 75*, 2867–2878.

18. Biswas, K., Chattopadhyay, I., Banerjee, R. K., & Bandyopadhyay, U., (2002). Biological activities and medicinal properties of neem (*Azadirachta indica*) *Curr. Scu., 82*, 1336–1345.

19. NAS, (1992). *Neem, a Tree for Solving Global Problems*. National Academy of Sciences, Washington, DC., USA.

20. Dua, V. K., Nepal, B. N., & Sharma, V. P., (1995). Repellent action of neem cream against mosquitoes. *Indian J. Malariol., 32*, 47–55.

21. Thas, J. J., (2008). Siddha medicine: Background and principles and the application for skin diseases. *Clin. Dermatol., 26*, 62–78.

22. Kabeeruddin, H., & Makhzanulmufradat, (2007). New Delhi: Aijaz Publishing House (pp. 400–411).

23. Tandon, P., & Sirohi, A., (2010). Assessment of Larvicidal properties of aqueous extracts of four plants against *Culex quinquefasciatus* larvae. *Jordan Journal of Biological Sciences, 3*(1), 1–6.

24. Ghani, N., (2004). *Khazainul Advia* (pp. 1330–1334). New Delhi, Idara Kitabul Shifa (YNM).

25. Chatterjee, A., & Pakrashi, S. C., (2010). *The Treatise on Indian Medicinal Plants* (Vol. 3, pp. 75–78). New Delhi: National Institute of Science Communication (CSIR).

26. Aditi, G., Bhandari, B. S., & Rai, N., (2011). Antimicrobial activity of medicinal plants *Azadirachta indica A. juss, Allium cepa* L. and *Aloe Vera* L. *Int. J. Pharm Tech. Res., 3*(2), 1059–1065.

27. Dua, V. K., Pandey, A. C., Raghavendra, K., Gupta, A., Sharma, T., & Dash, A. P., (2009). Larvicidal activity of neem oil (*Azadirachta indica*) formulation against mosquitoes. *Malar. J., 8*(1), 2–7.

28. Shravan, K. D., Ramakrishna, R., Santhosh, K. M., & Kannappan, N., (2011). *In vivo* antidiabetic evaluation of neem leaf extract in alloxan-induced rats. *Journal of Applied Pharmaceutical Science, 1*(4), 100–105.

29. Sithisarn, P., Supabphol, R., & Gritsanapan, W., (2005). Antioxidant activity of Siamese neem tree (VP1209). *J. Ethnopharmacol., 99*(1), 109–112.

30. Bandyopadhyay, U., Biswas, K., Sengupta, A., Moitra, P., Dutta, P., Sarkar, D., et al., (2004). Clinical studies on the effect of Neem (*Azadirachta indica*) bark extract on gastric secretion and gastroduodenal ulcer. *Life Sci., 75*, 2867–2878.

31. Isah, A. B., Ibrahim, Y. K., & Iwalewa, E. O., (2003). Evaluation of the antimalarial properties and standardization of tablets of *Azadirachta indica* (Meliaceae) in mice. *Phytother. Res., 17*(7), 807–810.

32. Talwar, G. P., Raghuvanshi, P., Misra, R., Mukherjee, S., & Shah, S., (1997). Plant immunomodulators for termination of unwanted pregnancy and for contraception and reproductive health. *Immunol. Cell Biol., 75*(2), 190–192.

33. Joshi, S. N., Dutta, S., Kumar, B. K., Katti, U., Kulkarni, S., Risbud, A., & Mehendale, S., (2008). *Expanded Safety Study of Praneem Polyherbal Vaginal Tablet Among HIV-Uninfected Women in Pune, India: A Phase II Clinical Trial Report, 84*(5), 343–347.

34. Garg, S., Doncel, G., Chabra, S., Upadhyay, S. N., & Talwar, G. P., (1994). Synergistic spermicidal activity of neem seed extract, reetha saponins and quinine hydrochloride. *Contraception, 50*(2), 185–190.

35. Pai, M. R., Acharya, L. D., & Udupa, N., (2004). Evaluation of antiplaque activity of *Azadirachta indica* leaf extract gel—a 6-week clinical study. *J. Ethnopharmacol., 90*(1), 99–103.

36. Vanka, A., Tandon, S., Rao, S. R., Udupa, N., & Ramkumar, P., (2001). The effect of indigenous neem *Azadirachta indica* mouthwashes on *Streptococcus mutans* and lactobacilli growth. *Indian J. Dent. Res., 12*(3), 133–144.

37. Chattopadhyay, R. R., Chattopadhyay, R. N., & Maitra, S. K., (2000). Effects of neem on hepatic glycogen in rats. *Indian J. Pharmacol., 25*, 174–175.

38. Gogate, S. S., & Marathe, A. D., (1989). *J. Res. Edu. Indian Med., 8*, 1.

39. Rao, A. R., Kumar, S. S., Paramasivam, T. B., Kamalkashi, S., Parashuram, A. R., & Shantha, M., (1969). *Indian J. Med. Res., 57*, 495.

40. Badam, L., Joshi, S. P., & Bedekar, S. S., (1999). *J. Commun. Dis., 31*, 79.

41. Parida, M. M., Upadhyay, C., Pandya, G., & Jana, A. M., (2002). *J. Ethnopharmacol., 79*, 273.

42. Board on Science and Technology for International Development, (1992). National Research Council, *Report of an Ad Hoc Panel on Neem, a Tree for Solving Global Problems* (pp. 60–113). National Academy Press, Washington D.C. (115).

43. Manikandan, P., Ramalingam, S. M., Vinothini, G., Ramamurthi, V. P., Singh, I. P., Anandan, R., et al., (2012). Investigation of the chemopreventive potential of neem leaf subfractions in the hamster buccal pouch model and phytochemical characterization. *Eur. J. Med. Chem. [Internet], 56*, 271–281. Available from: http://dx.doi.org/10.1016/j.ejmech.2012.08.008 (accessed on 29 June 2020).

44. Paul, R., Prasad, M., & Sah, N. K., (2011). Anticancer biology of *Azadirachta indica* L. (neem): A mini review. *Cancer Biol. Ther., 12*(6), 467–476.

45. SaiRam, M., Ilavazhagan, G., Sharma, S. K., Dhanraj, S. A., Suresh, B., Parida, M. M., et al., (2000). Anti-microbial activity of a new vaginal contraceptive NIM-76 from neem oil (*Azadirachta indica*). *J. Ethnopharmacol., 71*(3), 377–382.

CHAPTER 2

Arsenals of Pharmacotherapeutically Active Proteins and Peptides: Old Wine in a New Bottle

KIRTI DUBLI

Hislop School of Biotechnology, Hislop College, Nagpur – 440001, Maharashtra, India, E-mail: kirtidubli@gmail.com

2.1 INTRODUCTION

The word protein is derived from term *proteios* meaning either first or pre-eminent. This term was coined by Berzelius, a Swedish chemist, in 1838. Proteins are highly diverse molecules composed of amino acids. The world of peptides and proteins are fascinating. Proteins are building blocks of every organism including plants, bacteria, and fungi. Figure 2.1 indicates some proteins; for example, skin, eggs, hair, nails, etc., are all proteins. Not just at the macroscopic level but even major molecules like cell surface receptors, enzymes are peptides or proteins.

Egg Albumin

FIGURE 2.1 The example showing egg albumin as the most common protein form.

2.2 AMINO ACIDS

Amino acids can be simply referred to as monomers of proteins. The various amino acids are linked together to form peptides or proteins. The general structure of amino acids is as shown in Figure 2.2.

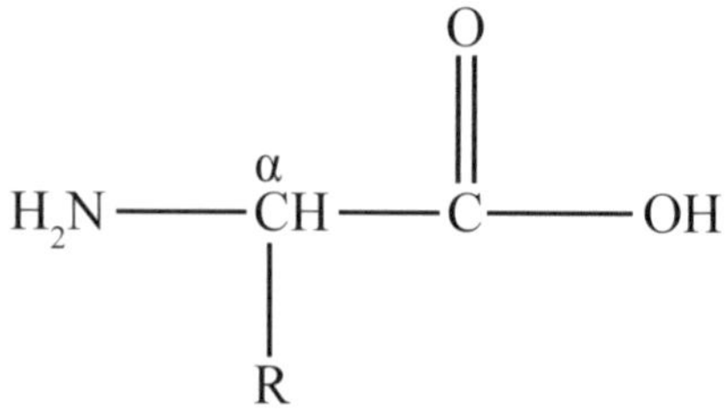

FIGURE 2.2 General structure of amino acids indicating the α-carbon.

Amino acid as the term indicates contains an amino group (-NH2) and a carboxylic acid group (–COOH) linked to the central carbon atom. The carbon atom to which both amino and the carboxylic acid group is attached is referred to as α-carbon. The fourth valence of the carbon is satisfied by a R. R can be hydrogen or methyl or any other group. The R group determines the amino acid. The proteins occurring in living organisms are primarily made of 20 amino acids. The name, structure, notation, and R group is mentioned in Table 2.1 [1].

TABLE 2.1 Name, Notation, and Structure of Amino Acids

Name of the Amino Acid	One Letter Notation	Three Letter Notation	Structure
Glycine	G	Gly	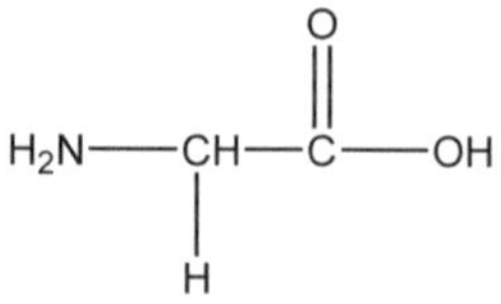
Alanine	A	Ala	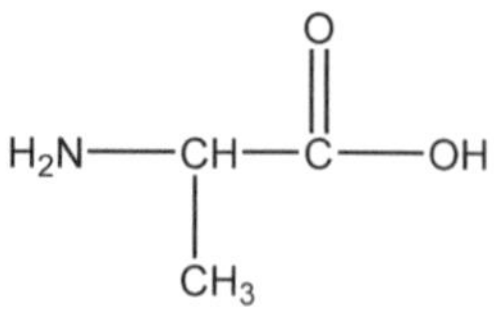

TABLE 2.1 *(Continued)*

Name of the Amino Acid	One Letter Notation	Three Letter Notation	Structure
Valine	V	Val	
Leucine	L	Leu	
Isoleucine	I	Ile	
Serine	S	Ser	
Threonine	T	Thr	

TABLE 2.1 *(Continued)*

Name of the Amino Acid	One Letter Notation	Three Letter Notation	Structure
Cysteine	C	Cys	H_2N—CH—C(=O)—OH, side chain CH_2—SH
Methionine	M	Met	H_2N—CH—C(=O)—OH, side chain CH_2—CH_2—S—CH_3
Aspartic acid	D	Asp	H_2N—CH—C(=O)—OH, side chain CH_2—C(=O)—OH
Glutamic acid	E	Glu	H_2N—CH—C(=O)—OH, side chain CH_2—CH_2—C(=O)—OH

TABLE 2.1   *(Continued)*

Name of the Amino Acid	One Letter Notation	Three Letter Notation	Structure
Asparagine	N	Asn	$H_2N-CH-C-OH$ with side chain CH_2, $C=O$, NH_2; carbonyl O
Glutamine	Q	Gln	$H_2N-CH-C-OH$ with side chain CH_2, CH_2, $C=O$, NH_2; carbonyl O
Lysine	K	Lys	$H_2N-CH-C-OH$ with side chain CH_2, CH_2, CH_2, CH_2, NH_2; carbonyl O

TABLE 2.1 *(Continued)*

Name of the Amino Acid	One Letter Notation	Three Letter Notation	Structure
Arginine	R	Arg	
Histidine	H	His	
Tryptophan	W	Trp	

TABLE 2.1 *(Continued)*

Name of the Amino Acid	One Letter Notation	Three Letter Notation	Structure
Proline	P	Pro	
Phenylalanine	F	Phe	
Tyrosine	Y	Tyr	

Selenocysteine (Sec) is debated to be the 21st amino acid shown in Figure 2.3. The thiol (-SH) group of cysteine is replaced by selenium atom (Se) to form the molecule of selenocysteine. Selenium functions as a trace element in human beings and is a part of selenoproteins. The pKa of selenocysteine is much lower than that cysteine and therefore, selenium atom of the seleno-cysteine is negatively charged and the thiol group is positively (protonated) at normal physiological pH.

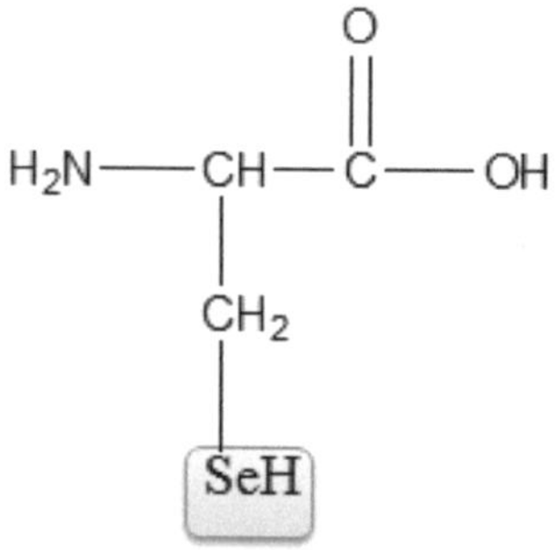

$$H_2N \longrightarrow CH \longrightarrow \overset{\overset{\displaystyle O}{\|}}{C} \longrightarrow OH$$

$$\underset{\underset{\boxed{SeH}}{|}}{\overset{|}{CH_2}}$$

FIGURE 2.3   Structure of selenocysteine.

Many proteins involved in redox reactions are selenoproteins [2]. For example, glutathione (GSH) reductases involved in the reduction of hydroperoxides [2], thioredoxin reductase functioning in NADPH dependent reduction of thioredoxin, iodothyronine deiodinase in thyroid hormone formation, selenoprotein W in muscle metabolism, sperm capsule selenoprotein required for motility of sperms [3].

Synthesis of selenoproteins comprises the inclusion of selenocysteine (Sec) in the proteins during translation. As all the 64 codons code for above mentioned 20 amino acids, it is very curious that what might be the codon for selenocysteine. Selenocysteine is included in the proteins when the codon UGA is not read as stop codon or the termination codon.

Selenium is obtained by the cell in the form of selenite or selenate. The selenite is reduced to selenide with the help of the GSH-glutaredoxin system and the thioredoxin system. However, the mechanism of conversion of selenite to selenide is not completely determined. Selenide is converted to monselenophosphate by seleno-phosphate synthetase 2 (SPS2). Monoselenophosphate functions as selenium donor for synthesis of Sec-tRNASec. Selenocysteine synthetase (SelA) aids in conversion of Ser-tRNASec to form Sec-tRNASec [2, 4]. Selenoprotein synthesizing mRNAs have a unique secondary structure in the 3'untranslated region, called SECIS element. The synthesis of selenoprotein also involves other molecules like Sec-specific elongation factor (SelB), Sec-tRNASec, SBP2 (SECIS-binding protein 2), ribosomal protein L30, 43-kDa RNA-binding protein, soluble liver antigen protein, and SPS1. All these molecules operate together to accomplish insertion Sec onto the growing polypeptide chain [4].

In *E. coli*, the SECIS element is positioned immediately after the UGA codon. SECIS element directs SelB to UGA codon for incorporation of Sec

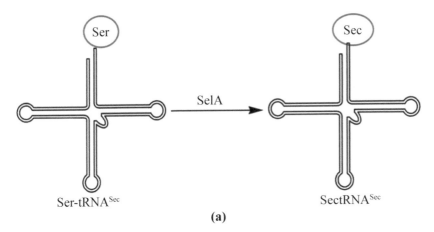

(a)

FIGURE 2.4 Mechanism of incorporation of selenocysteine in the selenoproteins during protein synthesis. (a) Conversion of Ser-tRNA^{Sec} to Sec-tRNA^{Sec} by selenocysteine synthetase (SelA).

in the polypeptide. SelB elongation factor has a SECIS binding domain at the C-terminal. SelB assists in the addition of SectRNASec to the growing polypeptide chain [2] as shown in Figure 2.4.

2.3   UNUSUAL AMINO ACIDS

Besides, there are also some unusual or non-standard amino acids occurring in living organisms which may or may not be a part of protein. Examples of such amino acids include 6-N-methyllysine, γ-carboxyglutamate, 4-hydroxyproline, 5-hydroxylysine, desmosine, ornithine, norvaline, etc. The structures of a few of these amino acids are shown in Figure 2.5.

6-N-methyllysine is obtained by transmethylation of lysine. Methylation of lysine is important in transcriptional regulation and double-strand break repair of DNA molecule in many organisms [7]. Post translation carboxylation of glutamic acid residues in proteins, results in γ-carboxyglutamate [8]. 4-hydroxyproline and 5-hydroxylysine, unusual amino acids, are an important part of collagen [6]. As these are not essential amino acids, they are synthesized by the organisms and not obtained from the diet. Desmosine is formed by four lysine molecules linked together forming a pyridinium ring. Desmosine helps in the cross-linking of elastin [9]. Ornithine is an important part of the urea cycle and the precursor of citrulline and arginine [10]. Norvaline is a component of antifungal peptides in bacteria. This unusual

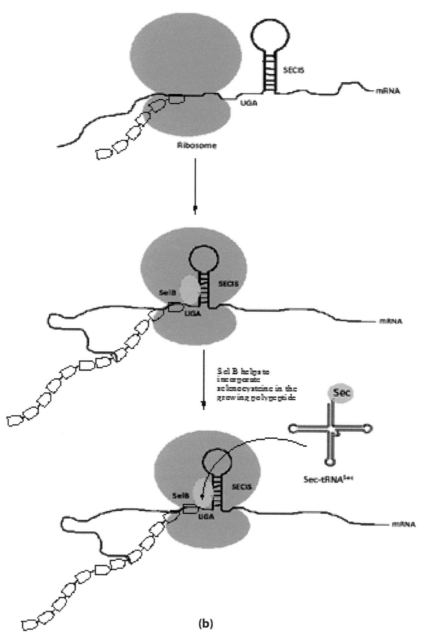

(b)

FIGURE 2.4 Mechanism of incorporation of selenocysteine in the selenoproteins during protein synthesis. (b) guiding of SelB to UGA codon by SECIS element and incorporation of selenocysteine in the growing polypeptide at UGA codon by Sec-tRNA^{Sec}.

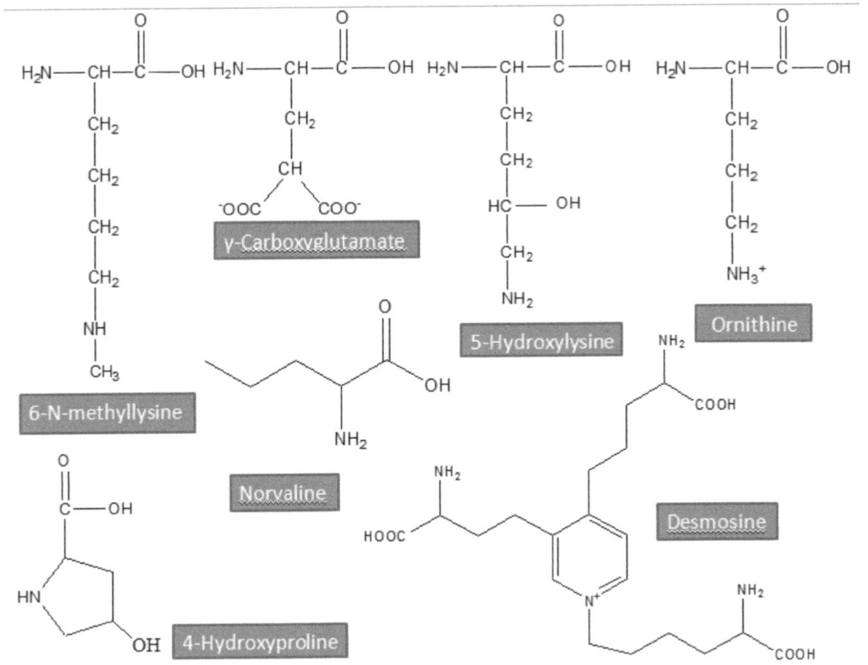

FIGURE 2.5 Structure of unusual amino acids.

amino acid is also found to be present in recombinant proteins synthesized using bacterial cells [11].

2.4 CLASSIFICATION OF AMINO ACIDS

The amino acids being an integral part of life are classified in various ways. A comprehensive approach of classification of amino acids is given as follows:

- Based on structure of R group;
- Based on polarity;
- Based on dietary needs;

2.4.1 CLASSIFICATION BASED ON STRUCTURE OF R GROUP

Based on the structure of R group, amino acid can be classified into seven different classes as shown in Table 2.2.

TABLE 2.2   Classification of Amino Acids Based on Structure of R Group

Structure of R Group	Amino Acids
Aliphatic	Glycine, Valine, Alanine, Leucine, and Isoleucine
Aromatic	Phenylalanine, Tryptophan, and Tyrosine
Sulfur-containing	Cysteine and Methionine
Imino acid	Proline
Neutral	Serine, Threonine, Asparagine, and Glutamine
Acidic	Aspartic acid and Glutamic acid
Basic	Lysine, Arginine, and Histidine

1. **Aliphatic Amino Acids:** The amino acids with an aliphatic R group include glycine, valine, alanine, leucine, and isoleucine [12].
2. **Aromatic Amino Acids:** Amino acids phenylalanine, tryptophan, and tyrosine are aromatic in nature. These amino acids contain a benzene ring. In tryptophan, an indole ring with a propanoic group is found [12, 13].
3. **Sulfur-containing Amino Acids:** Cysteine and methionine are two amino acids, which contain a sulfur atom in them. Methionine has a thiol group while the cysteine has a mercapto group [12].
4. **Imino Acid:** Proline is an imino acid and is a pyrrolidine derivative [12, 14].
5. **Neutral Amino Acids:** Serine, threonine, asparagine, and glutamine are neutral amino acids. Serine and threonine are neutral amino acids containing a hydroxyl group. Asparagine and Glutamine contain an amide in the R side chain [12].
6. **Acidic Amino Acids:** Aspartic acid and glutamic acid are amino acids containing the carboxyl acid group in the side chain [12].
7. **Basic Amino Acids:** Arginine, lysine, and histidine are basic amino acids. Lysine has an amino group and arginine has a guanidine group in the R side chain. Histidine has an imidazole ring in the R side chain (Figure 2.6) [12].

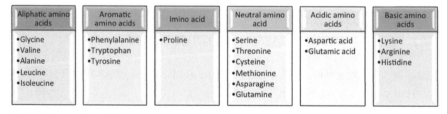

FIGURE 2.6   Classification based on structure of R group.

2.4.2   CLASSIFICATION BASED ON POLARITY

Polar molecules are those which have an affinity for water or are hydrophilic. Non-polar molecules are hydrophobic in nature. The classification of amino acids based on polarity is majorly governed by the type of R group present in the amino acid molecule. The nature of R group will determine the property of peptides and proteins at the biological pH. The amino acids are grouped as follows:

1.   **Non-Polar Amino Acids:** These amino acids have a hydrophobic or non-polar R group. The amino acids included in this group are alanine, valine, leucine, isoleucine, proline, phenylalanine, tryptophan, and methionine.
2.   **Uncharged Polar Amino Acids:** These amino acids have an uncharged polar or hydrophilic R group. The amino acids which can be categorized under this group are glycine, serine, threonine, tyrosine, cysteine, asparagine, and glutamine.
3.   **Negatively Charged Polar Amino Acids:** These amino acids bear a negative charge on their R groups at biological pH. Aspartic acid and glutamic acid are two such amino acids.
4.   **Positively Charged Polar Amino Acids:** These amino acids bear a positive charge on their R groups at biological pH. Lysine, Arginine, and Histidine are included in this group [1, 15].

The schematic representation of the classification is given in Figure 2.7.

2.4.3   CLASSIFICATION BASED ON REQUIREMENT IN DIET

Based on dietary needs, the amino acids are classified into three groups:

• **Essential Amino Acids:** The amino acids which cannot be synthesized by the human body and must be supplemented through diet are referred to as essential amino acids. For example, histidine, leucine, isoleucine, methionine, tryptophan, threonine, valine, lysine, and phenylalanine.
• **Non-Essential Amino Acids:** The amino acids which can be synthesized by the human body and need not be supplemented by the diet are referred to as non-essential amino acids. For example, alanine, aspartic acid, glutamic acid, and asparagine.
• **Conditional Amino Acids:** The amino acids which are required during the time of illness and stress but not usually essential are referred to as

conditional amino acids. For example, arginine, cysteine, glutamine, tyrosine, glycine, proline, and serine [15].

The schematic representation of the classification is given in Figure 2.8.

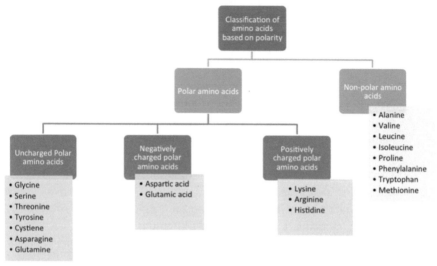

FIGURE 2.7   Classification of amino acids based on polarity.

Essential amino acids	Non-essential amino acids	Conditional amino acids
☐ Histidine	☐ Alanine	☐ Arginine
☐ Leucine	☐ Aspartic acid	☐ Cysteine
☐ Isoleucine	☐ Glutamic acid	☐ Glutamine
☐ Methionine	☐ Asparagine	☐ Tyrosine
☐ Tryptophan		☐ Glycine
☐ Threonine		☐ Proline
☐ Valine		☐ Serine
☐ Lysine		
☐ Phenylalanine		

FIGURE 2.8   Classification of amino acids based on requirements in diet.

2.4.4 *CLASSIFICATION BASED ON METABOLIC FATE AFTER DEGRADATION*

The amino acids can be broadly categorized into two groups based on the metabolic fate after their degradation. The classification is as follows:

1. **Ketogenic Amino Acids:** Amino acids which produce acetyl CoA or acetoacetyl CoA on degradation are referred to as ketogenic amino acids. Acetyl CoA or acetoacetyl CoA can combine to form ketone bodies or fatty acids and hence the name. For example, leucine and lysine.

2. **Glucogenic Amino Acids:** Amino acids which produce pyruvate, α-ketoglutarate, succinyl CoA, fumarate, or oxaloacetate on degradation are referred to as glucogenic amino acids. The products formed after degradation can be converted to phosphoenolpyruvate and then into glucose molecules. For example, glycine, valine, serine, cysteine, aspartate, glutamine, etc.

Not all amino acids can be classified as ketogenic or glucogenic. Some of them are both ketogenic and glucogenic in nature. For example, isoleucine, phenylalanine, tryptophan, and tyrosine [16].

2.5 PHYSICAL PROPERTIES OF AMINO ACIDS

The amino acids exhibit a wide range of physical properties. The melting point of amino acid tyrosine is 344°C while that of glutamic acid is 205°C. Most of the amino acids are colorless with the exception of tryptophan which is very pale yellow in color. The amino acids also possess varied tastes like sulfurous for methionine, umami for salts of glutamic acid, sweet for glycine. The physical properties of the amino acids are presented in Table 2.3.

2.6 OPTICAL PROPERTIES

The carbon atom adjacent to the carboxylic acid group is referred to as α-carbon. The α-carbon in all the amino acids is chiral with the exception of glycine. The orientation of $-NH_2$ group to the right or left of the α-carbon results in two enantiomers, D-form or L-form, respectively. Isoleucine and threonine have two chiral carbons and hence exist as 4 diastereomers ($2^2 = 4$). The other amino acids (except glycine) possess only one chiral carbon and therefore exist as 2 enantiomers ($2^1 = 2$) [18]. The D- and L-form of amino acids are indicated in Figure 2.9.

TABLE 2.3 Physical Properties of Amino Acids

Name of the Amino Acid	Melting Point (°C)	Boiling Point (°C)	Appearance	Density (g/cm³)	Odor	Taste	Solubility
Glycine	290	—	White crystals	1.61 at 20°C	Odorless	Sweet	Soluble in water; insoluble in ethanol and ethyl ether
Alanine	297	250	White Crystalline Powder	1.432 at 22°C	Odorless	Sweet	Soluble in water; insoluble in ether and acetone
Valine	315	Sublimes	White crystalline solid	1.23 at 25°C	—	—	Soluble in water; insoluble in neutral solvents
Leucine	293	Sublimes at 145–148°C	White crystals	1.293 at 18°C	—	—	Soluble in alcohol and water; insoluble in ether
Isoleucine	284	Sublimes at 168–170°C	Crystals	—	—	Bitter	Soluble in water; insoluble in ether
Serine	228	—	Colorless crystals	1.6 at 22°C	—	Sweet	Soluble in water; insoluble in benzene, ether, and ethanol
Threonine	256	—	Colorless crystals	—	—	—	Soluble in water; insoluble in chloroform, ethanol, and ethyl ether
Cysteine	240	—	Colorless crystals	—	—	—	Soluble in water, alcohol, acetic acid, and ammonia water; insoluble in ether, acetone, ethyl acetate, benzene, carbon disulfide, and carbon tetrachloride
Methionine	280–282	181	Colorless crystalline powder	—	Faint	Sulfurous	Soluble in water; insoluble in alcohol, ether, petroleum ether, benzene, and acetone
Aspartic acid	270–271	—	White crystalline solid	1.6603 at 13°C	—	Sour	Soluble in water; insoluble in benzene, ethyl ether and ethanol
Glutamic acid	205	—	White crystalline powder	1.525	—	Umami (salt of glutamic acid)	Soluble in water

TABLE 2.3 *(Continued)*

Name of the Amino Acid	Melting Point (°C)	Boiling Point (°C)	Appearance	Density (g/cm³)	Odor	Taste	Solubility
Asparagine	234–235	—	Orthorhombic bisphenoidal crystals	1.543 at 15/4°C	—	Neutral (L-isomer); Sweet (D-isomer)	Soluble in water, acids, and alkalis; insoluble in methanol, ethanol, ether, and benzene
Glutamine	185.5	—	Dry powder	1.364	—	—	Soluble in water
Lysine	224.5	—	Colorless crystals	—	—	Sweet/Bitter	Soluble in water; insoluble in ethanol, ethyl ether, acetone, and benzene
Arginine	244	—	Prisms containing 2 mol H_2O from water; anhydrous monoclinic plates from 66% alcohol	—	—	—	Soluble in water; insoluble in ethyl ether
Histidine	287	—	Colorless; Needles or plates	—	—	Sweet	Soluble in water; insoluble in ethyl ether and acetone
Tryptophan	282	—	White to yellowish-white crystals	—	Odorless	Slightly bitter	Soluble in water; insoluble in ethyl ether
Proline	220–222	—	Colorless crystals	—	—	Sweet/Bitter	Soluble in water; insoluble in ether, butanol, and isopropanol
Phenylalanine	283	—	Prims from water	—	Slight	Bitter	Soluble in water; insoluble in ethanol, ethyl ether, and benzene
Tyrosine	344	Sublimes	White crystals	—	—	—	Soluble in water; insoluble in absolute alcohol, ether, and acetone

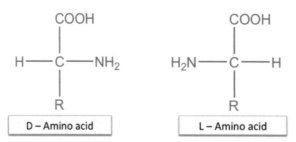

FIGURE 2.9   D-form and L-form of amino acids.

The L-form of amino acids is a naturally occurring form of amino acids. The L-amino acids found in the natural proteins are laevorotatory as well as dextrorotatory. The values of specific rotation for L-amino acids are indicated in Table 2.4. The D-forms of amino acids were reported in bacteria *Bacillus brevis, Lactobacillus arabinosus,* etc. [17]. The peptidoglycan of bacteria contains D-Ala and D-Glu. The bacterial peptides like Gramicidine S (D-Phe is present), Gramicidine D (D-Leu, D-Val, and D-Phe), etc., contain D-amino acids. D-Amino acid oxidase (DAO) is an enzyme found in humans for catabolizing D-amino acids. It brings about oxidative deamination of D-amino acids in the body, i.e., D-Ala from bacteria of the small intestine, D-Ser, endogenous co-agonist of N-methyl D-aspartate glutamate receptors [19, 20].

TABLE 2.4   Specific Rotation of Amino Acids

Name of the Dextrorotatory Amino Acid	Specific Rotation	Name of the Laevorotatory Amino Acid	Specific Rotation
L-Alanine	+2.8° at 25°C/D water, Concentration = 6%)	L-Leucine	−10.8° at 25°C/D (c = 2.2)
L-Valine	+22.9° at 23°C/D (c = 0.8 in 20% HCl)	L-Serine	−6.83° at 20°C/D (1.5 G in 15 G aqueous solution)
L-Isoleucine	+40.6° at 20°C/D (c = 4.6 in 6.1 N HCl)	L-Aspartic acid	−2.0° at 25°C (c = 3.93 in 5N HCl)
L-cysteine	+6.5° at 25°C/D (5 N HCl)	L-Asparagine	−5.3° at 20°C (c = 1.3 mg/L)
L-Arginine	+26.9° at 20°C/C (c = 1.65 in 6.0 N HCl)	L-Histidine	−10.9° at 20°C/D (c = 0.77 in 0.5N of NaOH)
		L-Tryptophan	−31.5° at 23°C/D (water, 1%)
		L-Proline	−80.9° at 20°C/D (Water, 1%)
		L-Phenylalanine	−35.1° at 20°C/D (c = 1.94)
		L-Tyrosine	−10.6° at 22°C/D (c = 4 in HCl)

2.7   ISOELECTRIC POINT AND TITRATION CURVES

Amino acid as the name suggests contains an amino group and a carboxyl group. In cells, the amino acids are present in charged form, bearing a positive charge on the amino group and a negative charge on the carboxyl group. This arises because of proton transfer from the carboxyl group to the amino group. Thus, amino acid molecule bearing both the charges can act as an acid as well as base, and hence it is amphoteric. The form of amino acid in aqueous solutions or cells, at physiological pH, bearing a single positive charge and a single negative charge is referred to as zwitterion as shown in Figure 2.10.

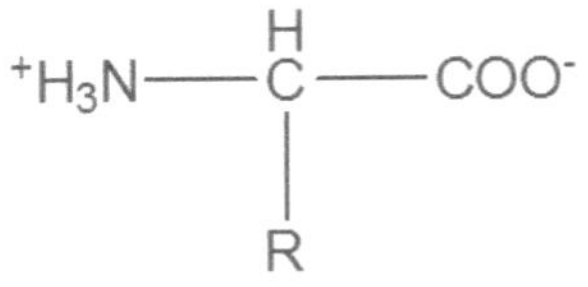

$$^+H_3N—\overset{\overset{\displaystyle H}{|}}{\underset{\underset{\displaystyle R}{|}}{C}}—COO^-$$

FIGURE 2.10   Zwitter ion.

At acidic pH, the amino acid molecule is completely protonated and bears a net positive charge (shown in Figure 2.11). At basic pH, the amino acid molecule is completely deprotonated and has a net negative charge (shown in Figure 2.12) [1, 22].

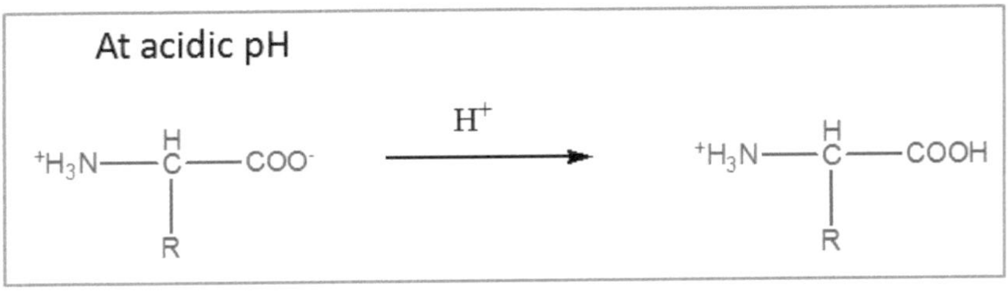

At acidic pH

$$^+H_3N—\overset{\overset{\displaystyle H}{|}}{\underset{\underset{\displaystyle R}{|}}{C}}—COO^- \quad \xrightarrow{H^+} \quad ^+H_3N—\overset{\overset{\displaystyle H}{|}}{\underset{\underset{\displaystyle R}{|}}{C}}—COOH$$

FIGURE 2.11   Amino acid under acidic condition.

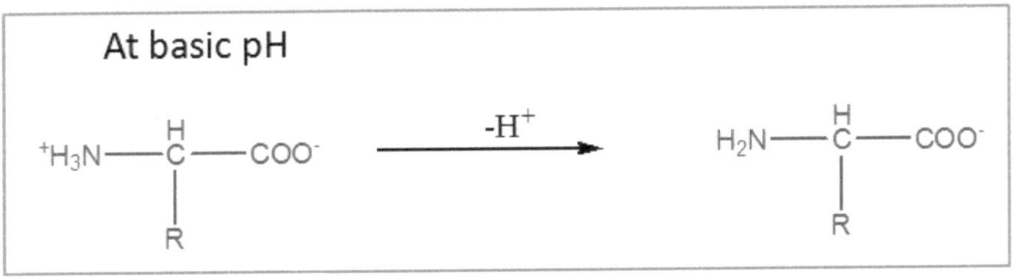

At basic pH

$$^+H_3N—\overset{\overset{\displaystyle H}{|}}{\underset{\underset{\displaystyle R}{|}}{C}}—COO^- \quad \xrightarrow{-H^+} \quad H_2N—\overset{\overset{\displaystyle H}{|}}{\underset{\underset{\displaystyle R}{|}}{C}}—COO^-$$

FIGURE 2.12   Amino acid under basic environment.

The behavior of amino acids at different pH can be more easily understood from the titration curves. The amino acid titration involves gradual addition or removal of proton. For obtaining the titration curve, the amino acid is titrated against a strong base like sodium hydroxide (NaOH). The titration curve of alanine is shown below in Figure 2.13.

At low pH, the molecule of alanine bears a positive charge due to protonation ($^+H_3N–CH(CH_3)–COOH$). As a strong base is added to it gradually, alanine loses its protons step by step. The proton of the carboxyl group is lost first because of the low pK_a value. As the titration proceeds, a stage is reached where the net charge on the molecule of alanine is neutral (a positively charged $–NH^{3+}$ group and a negatively charged $–COO^-$ group on each molecule). The pH at which the amino acid molecule bears a neutral charge is referred to as isoelectric point, pI. Further addition of base results in loss of proton from ammonium group. Thus, an amino group is formed and the amino acid molecule bears a net negative charge ($H_2N–CH(CH_3)–COO^-$) [21]. The titration curve of aspartic acid (Figure 2.14) and lysine are shown in Figure 2.15.

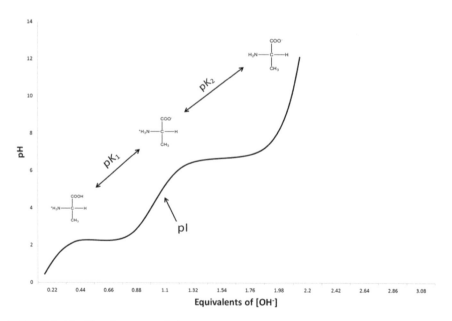

FIGURE 2.13   Titration curve of alanine.

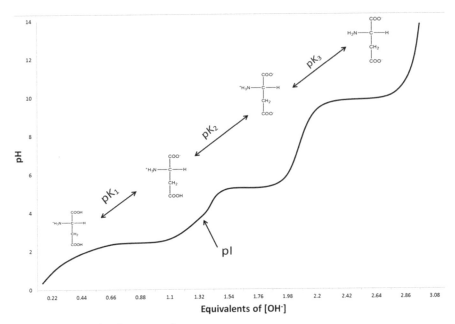

FIGURE 2.14 Titration curve of aspartate.

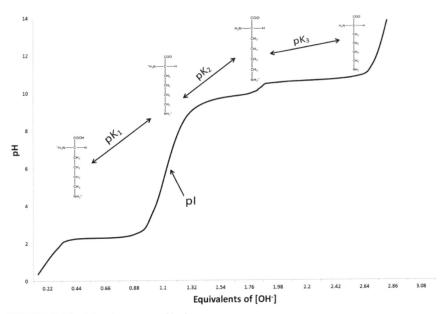

FIGURE 2.15 Titration curve of lysine.

Alanine has two pK_a values, $pK_1 = 2.34$ and $pK_2 = 9.69$ corresponding to the carboxyl group and amino group respectively. These pK values are the pH at which equilibrium of conversion of protonated amino acid to zwitterion and conversion of zwitterion to negatively charged amino acid is achieved. The isoelectric point, pI = 6.01, is an average of pK_1 and pK_{a2} [22].

$$pI = \frac{pK_1 - pK_2}{2}$$

A third dissociation constant is given by pK_R, in case the amino acid has a dissociable R group [23]. The dissociation constants and pI of different amino acids are given in Table 2.5.

The titration curves of amino acid facilitate the understanding of buffering capacity of the amino acid. The titration curve of alanine indicates that it has buffering capacity at pH of 2.34 and 9.69 while the least buffering capacity at its isoelectric point 6.01.

The isoelectric point of a protein is that pH at which it bears no net charge. The isoelectric point of proteins cannot be deduced from its titration curve as it is very complex. This complexity arises because of large no of groups which can be ionized, protein folding leading to its 3-dimensional structure and interaction of different groups within the protein [23].

TABLE 2.5 Dissociation Constants and Isoelectric Point of Amino Acids

Name of the Amino Acid	pK_1 (–COOH)	pK_2 (-NH^{3+})	pK_R	pI
Glycine	2.34	9.6	–	5.97
Alanine	2.34	9.69	–	6.01
Valine*	2.32	9.62	–	5.97
Leucine	2.36	9.63	–	5.99
Isoleucine	2.36	9.68	–	6.02
Serine*	2.21	9.15	13.60	5.68
Threonine*	2.11	9.62	13.60	5.87
Cysteine	1.71	8.33	10.78	5.02
Methionine	2.28	9.21		5.74
Aspartic acid	1.92	9.87	3.87	2.89
Glutamic acid*	2.19	9.67	4.25	3.22
Asparagine	2.02	8.80		5.41
Glutamine*	2.17	9.13		5.65

TABLE 2.5 *(Continued)*

Name of the Amino Acid	pK$_1$ (–COOH)	pK$_2$ (-NH^{3+})	pK$_R$	pI
Lysine	2.18	8.95	10.53	9.74
Arginine*	2.17	9.04	12.48	10.76
Histidine*	1.82	9.17	6.00	7.59
Tryptophan	2.38	9.39		5.89
Proline	1.99	10.60		6.30
Phenylalanine	1.83	9.13		5.48
Tyrosine	2.20	9.11	10.07	5.66

2.8 BIOLOGICAL ROLES OF SOME AMINO ACIDS

Besides being part of protein, molecule amino acids also perform some unique functions. Some of these are described briefly below.

Glycine plays a significant role in central nervous system. It functions as an inhibitory neurotransmitter and also as co-agonist with glutamate in activation of N-methyl-D-aspartate receptors. It also functions as an inhibitory neurotransmitter along with γ-amino butyric acid (GABA) [24].

Lysine is required for synthesis of collagen, elastin, and carnitine [25]. Arginine is an important amino acid required for synthesis of molecules like nitric oxide, urea, creatine, proline, polyamine, and glutamate. The presence of arginine is also required for expression of some genes like arginosuccinate synthetase, arginosuccinate lyase, iNOS, cationinc amino acid transporter-1 (CAT-1) and zeta chain of T-cell receptor [26]. Histamine, a histidine derivative, plays an important role in inflammation. It also functions in gastric acid secretion and neuromodulation [34].

Aspartame an ordinarily available sweetener is obtained from the amino acids aspartic acid and phenylalanine. This artificial sweetener is 200 times sweeter than sugar [27]. The metabolism of aspartame produces aspartic acid, phenylalanine, and methanol. Various studies have been done to analyze the toxicity of aspartame. It was found that the aspartame related damage results when the amount consumed is much higher than the human consumption [28]. Glutamate functions as an excitatory neurotransmitter, besides functioning in transamination reaction, which helps in maintaining the nitrogen levels in the body. Ammonia is stored and transported in the form of glutamine through the blood. It also functions as precursor in GABA, purine, and pyrimidine synthesis [29].

Cysteine, a non-essential sulfur-containing amino acid, has antioxidant features and is required for production GSH (a tripeptide composed of glutamate, glycine, and cysteine). GSH helps in regulation of action of vitamin C and vitamin E. Interestingly cysteine also helps in moderating the effects of hangover [30]. S-adenosyl methionine (SAM), a methionine derivative is essential for metabolism of polyamines and nucleic acids, precursor of GSH, and structure and function of plasma membrane.

Tryptophan is precursor of amino acid for serotonin (5-hydroxytryptamine). Serotonin is responsible for appetitive, emotional, motor, cognitive, and autonomic behaviors. Serotonin is also involved in the neuroendocrine function and circadian clock rhythm. It also functions as precursor of melatonin, which possess sedative and hypnotic properties [31, 32]. Tryptophan is also involved synthesis of molecules like niacin, tryptamine, and kynurenine. Dopamine, norepinephrine, and epinephrine are tyrosine derivatives. These tyrosine derivatives are neurotransmitters mediating various function of the nervous system [33].The structures of various amino acid derivatives are shown in Figure 2.16.

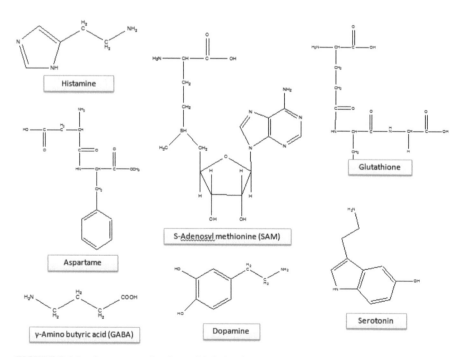

FIGURE 2.16   Structures of amino acid derivatives.

2.9   PEPTIDES

The amino acids link together and form peptides or proteins. The peptide bond covalently links the amino acid forming the backbone of the peptide chain. These peptides perform many important biological roles. They act anti-microbials, hormones, growth factors, etc.

2.9.1   *PEPTIDE BOND*

The amino acids are linked together to each other forming peptides and proteins. The bond which links different amino acids in a peptide chain is the peptide bond (–CONH-). The peptide bond is a covalent bond formed by condensation of –COOH bond of one amino acid and –NH_2 bond of another amino acid with the removal of a single water molecule (as shown in Figure 2.17). The resulting molecule is a peptide. The amino acids in a peptide are referred to as amino acid residue. When many amino acids are linked using peptide bonds, the molecule so formed is the peptide chain. The peptide bond has a length of 1.32Å [1, 35].

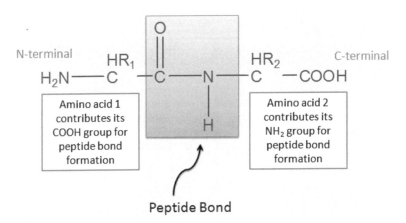

FIGURE 2.17   A dipeptide showing peptide bond with N- and C-terminal amino acids.

2.10   ISOPEPTIDE BOND AND ITS BIOLOGICAL ROLE

The isopeptide bond is a bond formed between amino groups of lysine residue of a protein with the carboxyl group of another protein molecule. The isopeptide bond is formed when the protein is tagged for degradation with ubiquitin. The ε amino group of lysine is linked to the carboxy-terminal of the ubiquitin. Thus, the ubiquitinylated protein is degraded by proteasome.

The isopeptide bonds are also observed in blood clots. The fibrin molecules are crosslinked by isopeptide bond between glutamine residues and ε amino group of lysine of fibrin, forming large polymeric fibrin network which prevents blood flow [36].

On the basis of number of amino acids present, peptides can be classified as:

1. Dipeptide which contains two amino acid residues.
2. Tripeptide which contains three amino acid residues.
3. Tetrapeptide which contains four amino acid residues.
4. Oligopeptides which contain less than 10 amino acid residues.
5. Polypeptides which contain 50 or less than 50 amino acid residues.

Polypeptides containing more than 50 amino acid residues are called proteins [1]. The polypeptides have an N-terminal and a C-terminal. The amino acid present at the N-terminal contributes –COOH to the peptide bond formation while the amino acid present at the C-terminal contributes –NH$_2$ to the peptide bond formation (as shown in Figure 2.18).

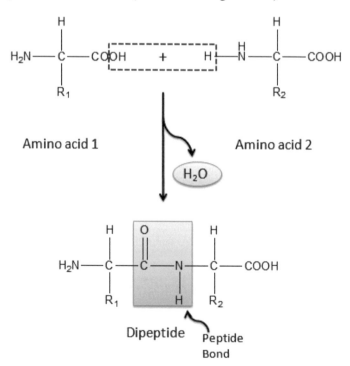

FIGURE 2.18   Formation of peptide bond.

2.11 NOMENCLATURE OF THE PEPTIDE CHAIN

The peptide is named starting from the amino acid present at its N-terminal, sequentially followed by the amino acid residues present in the peptide chain and the C-terminal amino acid in the end. The amino acid, with its conformation, at the N-terminal, and in the chain is written by replacing its suffix-ine/-ic acid with suffix-yl, while the name of amino acid at its C-terminal are written per se. For example, a peptide containing L-form of amino acid lysine, histidine, glycine, serine, and cysteine (from N-terminal to C-terminal) is named as L-Lysyl-L-histidyl-L-glycyl-L-seryl-L–cysteine. The amino acids present in the peptide chain can be represented by using a three-letter symbol or one letter symbol starting from N-terminal to C-terminal [37]. The above-mentioned peptide L-Lysyl-L-histidyl-L-glycyl-L-seryl-L-cysteine can, therefore, be represented as:

Lys-His-Gly-Ser-Cys

or

K-H-G-S-C

2.12 PROPERTIES OF PEPTIDE BOND

1. **Peptide Bond has a Partial Double Bond Character:** The peptide bond has a length of 1.32Å intermediate of C-N and C = N. The C-N has a length of 1.49Å and C = N has a length of 1.27Å. The electron pair presents on the nitrogen atom of the peptide bond can be donated to the oxygen of peptide bond resulting in partial double bond formation (Figure 2.19).

2. **Peptide Bond is a Planar Bond:** as the peptide bond has a partial double bond character, this limits the rotation about it. Thus, the six atoms, i.e., C_α, C, O, N, H, and C_α of the two adjacent amino acids linked by peptide bond lie in the same plane. The planarity of the peptide bond restricts the conformations of the peptide backbone.

3. **Peptide Bond is a Rigid Bond:** The partial double bond characteristic also makes the peptide bond a rigid bond. The angle of rotation (dihedral angle) between N atom and the C_α of the one amino acid residue in the peptide bond are denoted by φ and the angle of rotation (dihedral angle) between C atom and the C_α of the other amino acid residue in the peptide bond is denoted ψ. The C_α linked to N atom and C_α linked to C atom of the peptide bond is single bonded. The

peptide or protein structure cannot have all the values of φ and ψ because of steric hindrances (Figure 2.20).

a)

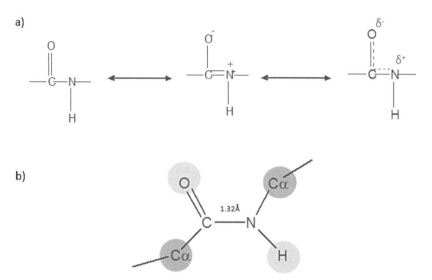

b)

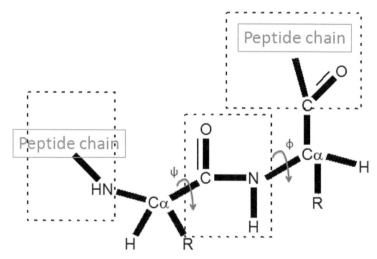

FIGURE 2.19   Properties of a peptide bond. (a) Partial double bond nature of peptide bond is indicated by resonance structure. (b) C-N bond length in peptide bond is 1.32Å, which is intermediate of C-N single bond and C = N double bond. The *trans*-configuration of the peptide bond is indicated by opposite orientation of Cα about the peptide bond.

FIGURE 2.20   The Peptide bond in a peptide.

Note: The hashed squares indicate the plane of the peptide bond. The φ indicates the rotation of N–Cα bond and ψ indicates the rotation of C–Cα bond.

4. **Peptide Bond can have a *cis* or *Trans*–Configuration:** When the C_α of the peptide bond is present on the same side, the bond has *cis* configuration and when they are present on the opposite side, the bond has *trans* configuration. Peptides and proteins generally have *trans* configuration about the peptide bond. The *cis* configuration is observed about the proline residues, at the turn in the peptide or protein molecule [36, 38].

2.13   FUNCTIONS OF PEPTIDES

Peptides are components of protein molecules. Moreover, peptides also perform some essential tasks which are described as follows:

1. **Antimicrobial Peptides:** These are synthesized by both plants and animals as an essential constituent of immune system. These peptides generally destroy the pathogen by membrane permeabilization, inhibition of synthesis of membrane proteins, inhibition of enzymes in pathogens, synthesis of stress proteins, breaking down ssDNA, production of hydrogen peroxide, etc. [40]. Some of the antimicrobial peptides in humans are given in Table 2.6.

TABLE 2.6   Role of Human Antimicrobials

Name of the Peptide	Mechanism of Action	Active Against
Histatins	Inducing cell death by non-lytic loss of ATP, disrupt cell cycle and generate reactive oxygen species (ROS)	Fungi
Defensins	Formation of pores on target cells	Bacteria, Fungi, and Enveloped Viruses
Cathelicidin LL-37	Attracts mast cells, monocytes, T-cells, and neutrophils at the site of injury/wound/infection	Bacteria, Fungi, and Viruses

2. **Opioid Peptides:** These peptides influence the nervous system by acting as hormone as well as neuromodulators. These peptides bring about various changes like inducing euphoria and analgesia, decreased respiration, movement of muscles of gastrointestinal (GI) tract and susceptibility to seizures, Incite or dampen cardiovascular

function, etc. [41]. Endorphins, Enkephalins, Dynorphins, and Endo-morphins are opioid peptides produced in the body [42].

3. **Peptide Hormones:** These are involved in regulation of many functions in the human body. They shall be described in the following sections in the chapter.

4. **Peptide Growth Factors:** Growth factors are responsible for stimulating cell division and cell differentiation. For example, epidermal growth factor (EGF) is required for mammary gland growth and differentiation of cells of small intestine in newborns [43], nerve growth factor is necessary for development and maintenance of nerve cells of peripheral nervous system [30].

5. Peptide like GSH (tripeptide) behaves as an antioxidant protecting the cell from free radical damage [44].

2.14   CYCLIC PEPTIDES

Cyclic peptides are those peptides which have ring structure in the molecule. The ring structure is formed by bonds like amide, ether, thioether, lactone, disulfide, etc. Many naturally occurring cyclic peptides have clinical uses. Some of these peptides are listed in Table 2.7 with their functions and their structures are given in Figure 2.21 [44].

2.15   STRUCTURE OF PROTEINS

The amino acids are monomers of proteins. Proteins are not simply assembled in a cell using peptide bond linking amino acids, but there are other bonds also involved in forming an efficient protein [1]. The bonds involved in forming a protein molecule are described below.

2.15.1   BONDS COMPRISING THE PROTEIN STRUCTURE

1. **Peptide Bond:** These are covalent bonds which link amino acids together to form a peptide or a protein. The peptide bond forms the backbone of the peptide or the protein. A peptide bond is formed by condensation reaction between –COOH of one amino acid and –NH_2 of another amino acid with release of water molecule. The peptide bond has a partial double bond character which provides rigidity as well as planarity to the peptide bond.

TABLE 2.7   Cyclic Peptides and Their Functions

Name of the Peptide	Functions
Tyrocidine	Bactericidal
Cyclosporin A	Immunosuppressant
RGD Peptide	Inhibits tumor cell growth; induces apoptosis of angiogenic blood vessels
Oxytocin	Stimulates contraction of smooth muscles of uterus at the time of parturition, contraction of mammary gland for milk ejection

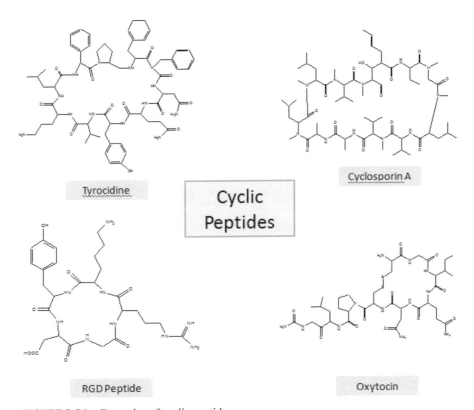

FIGURE 2.21   Examples of cyclic peptides.

2. **Disulfide Bond:** These are also covalent bonds responsible for forming the protein structure. The disulfide bond is formed between sulfur atoms of cysteine residues of the peptide(s) or protein(s). Hence, the disulfide bond formed may be intramolecular or intermolecular in nature. The formation of disulfide bond is shown in Figure 2.22.

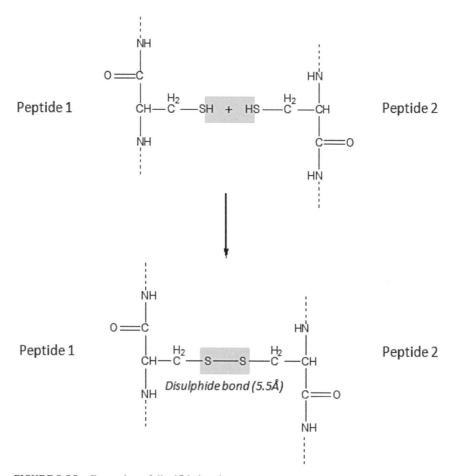

FIGURE 2.22　Formation of disulfide bond.

The two cysteine residues of a disulfide bonds are 5.5Å away from each other and are generally around 25 amino acid residues far apart on the protein molecule. The formation of a disulfide linkage in a protein is dependent upon the presence of oxidizing agent or reducing agent and the redox potential required for bond formation. Thus, disulfide bond formation regulates biological activity of the proteins or enzymes involved in the electron transfer processes.

The formation of disulfide bond of extracellular proteins occurs in the endoplasmic reticulum (ER) of eukaryotes. In gram-negative bacteria, the formation of disulfide bond is mediated by oxidases or isomerases in the periplasm [36].

3. **Hydrogen Bond:** It is a weak bond formed between hydrogen atoms bound to an electronegative atom, with another electronegative atom. The hydrogen bonds (H-bonds) in proteins are formed between hydrogens of amide and oxygen carbonyl groups of the peptide bonds in protein. The H-bonds are also formed between hydroxyl, amine, amide, and carboxyl groups of amino acid residues. The H-bonds have a length less than 2.5Å and bond angle between 90° and 180° between the bond forming moieties. H-bonds have strength of 10–40 kJmol^{-1} and play an important role in secondary structure of protein. The property of molecular recognition is bestowed upon the protein by the directionality and specificity of interaction between protein and the ligand by the hydrogen bond [45].

4. **Hydrophobic Interactions:** A protein molecule contains polar as well as non-polar amino acid residues. The non-polar amino acids like alanine, valine, phenylalanine, tryptophan, etc., contain hydrophobic R group. During protein folding these groups, congregate together forming the core of protein, reducing or completely diminishing any contact with water. This arrangement is energetically favorable and provides sufficient amount of free energy for maintaining the protein structure [46].

5. **Salt Bridges:** They are also referred to as ionic bonds or electrostatic bonds. These bonds are formed between positively charged amino acids (aspartic acid and glutamic acid) and negatively charged amino acids (Lysine, arginine, and histidine) at physiological pH. The formation of salt bridge limits the low energy conformations adopted by the protein molecule and therefore plays an important role in the specificity of the molecule [47, 48].

2.16 RAMACHANDRAN PLOT

The Ramachandran plot is an instrumental plot determining the structure of protein or peptides [49]. This plot deals with possible dihedral angles, φ and ψ, that can be taken up by the amino acid residues of peptide bonds in peptides or proteins. Ramachandran and his colleagues studied φ and ψ taken up by alanyl dipeptide mimetic, i.e., N-acetyl-L alanine methylester [49], by considering the atoms to be rigid spheres. There are three allowed conformations which are as follows:

1. Right-handed α helix;
2. β sheet (parallel and antiparallel); and

3.   Left-handed α helix.

These conformations are obtained on the basis of consideration that atoms have fixed radii and do not collide. Besides, the plot also shows the presence of regions called bridge regions which connect right-handed α helix and β sheet based on consideration that atoms with smaller radii can have the least possible value [50]. The Ramachandran plot has been shown in Figure 2.23.

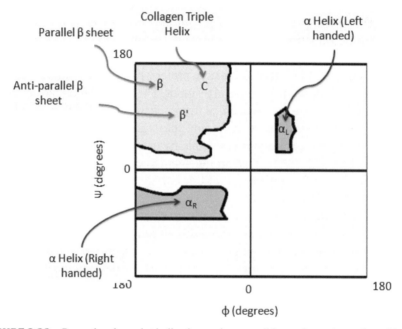

FIGURE 2.23   Ramachandran plot indicating various possible conformations of peptide.

Furthermore, Ramachandran and his colleagues studied the effect of back-bone angle τ on the various allowed combinations of φ and ψ. It was found that the bridge region is highly sensitive to τ, $\varphi < 0°$ and $-20° < \psi < 40°$ [49].

Glycine has hydrogen present as its R group and therefore causes very little steric hindrance in polypeptide when compared to other amino acid residues, thus allowing the polypeptide to turn. Glycine shows all the allowed conformations on the Ramachandran plot but these conformations occupy large regions [1]. Proline has a confined Ramachandran plot due to the 5-membered pyrrolidine ring. The pyrrolidine ring restricts the rotation of N–C$_\alpha$ bond. Due to these limitations, the φ is set about-65° [1, 51]. Hence,

proline generally occurs in the terminal regions of the polypeptide and inter-
rupts the helix [52].

2.17 HIERARCHY OF PROTEIN STRUCTURE

The various levels of organization of protein structure given by Linderstörm-
Lang are as follows:

1. Primary structure of protein;
2. Secondary structure of protein;
3. Tertiary structure of protein;
4. Quaternary structure of protein.

The primary, secondary, and tertiary structure of proteins is present in
the same protein molecule. The quaternary structure of protein deals with
interactions among different subunits of a protein. Each of the levels of the
protein structure is described below in detail [1].

2.17.1 PRIMARY STRUCTURE OF PROTEIN

The exact number and sequence of amino acid in a polypeptide or protein
molecule refers to the primary structure of protein. The sequence of amino
acid in a protein is governed by the genes. The significance of sequence of
amino acid in a protein is as follows:

1. Determining mechanism of action of the protein;
2. Establishing the three-dimensional structure of protein;
3. Studying mutations which lead to changes in protein and hence
   medical disorders;
4. Understanding evolutionary hierarchy.

The primary structure of some proteins contains disulfide bonds. For
example, a chain of insulin contains an intramolecular disulfide bond as
shown in Figure 2.24 [53].

2.17.2 SECONDARY STRUCTURE OF PROTEINS

The H-bonds form the basis of both α helix and β sheet [54]. The H-bonds
are formed between the Nitrogen atom of –NH group of one amino acid

residue and oxygen atom of –C = O group of another amino acid residue [55]. The various secondary structures of protein are α-helix, β-pleated sheet, and β turns.

FIGURE 2.24   Disulfide bond in insulin.

The structure of alpha-helix and beta-sheet was elucidated by Linus Pauling, Robert Corey, and Herman Branson in the year 1951. Pauling and Corey approach for deducing the structures was espoused by Watson and Crick for the determining structure of DNA molecule [54].

2.17.2.1   α-HELIX

α-helix is a right-handed helix with 3.69 amino acid residues per turn of the helix. The turn of the helix has a pitch of 5.4Å with every amino acid residue repeating 1.47Å. The amino acid residues in the helix are hydrogen-bonded to every fourth amino acid residue in the peptide chain. The distance between hydrogen-bonded nitrogen and oxygen is 2.72Å. The R groups of amino acid residues are oriented outwards from the axis of the helix [55]. The structure of the helix is represented in the figure. α-keratin of hair and wool possess α-helical arrangement [55]. The structure of α-helix is represented in Figure 2.25.

2.17.2.2   ANOTHER HELIX

Γ-helix has 5.1 residues amino acid residues per turn of the helix. The turn of the helix has a pitch of 5.03Å with every amino acid residue repeating 0.99Å. The amino acid residues in the helix are hydrogen-bonded to every sixth amino acid residue in the peptide chain [54, 55].

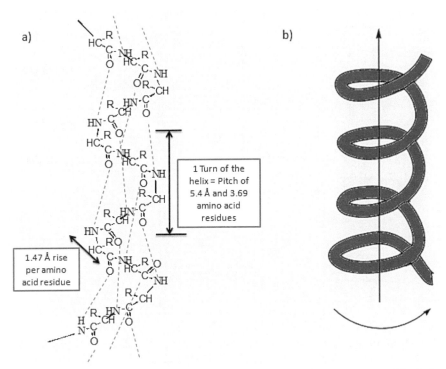

FIGURE 2.25 Structure of right-handed α-helix. (a) Schematic arrangement of the helix indicating the hydrogen bonds and various aspects of the helix; (b) right-handed helix.

2.17.2.3   β-PLEATED SHEET

β-pleated sheet is obtained by extended chain configuration of the peptide chains in a single plane. The H-bonds in the β-pleated configuration are formed between the peptide chains. The pleated sheet may be parallel or antiparallel in its disposition. In the parallel configuration of β-sheet, all the amino acid residues are in the same orientation, i.e., carbonyl groups of all the peptide chains are positioned in one direction while the amino group is positioned in the opposite direction. In the case of antiparallel β-sheet, the carbonyl and amino groups of the polypeptide chains are in the same direction [56]. The parallel and antiparallel forms of β-pleated sheet are represented in the figure. The β-pleated sheet arrangement can be within the same polypeptide chain or between two polypeptide chains [53] as shown in the diagram. The silk fibroin has β-pleated sheet arrangement. The structure of parallel and anti-parallel β sheet is shown in Figure 2.26.

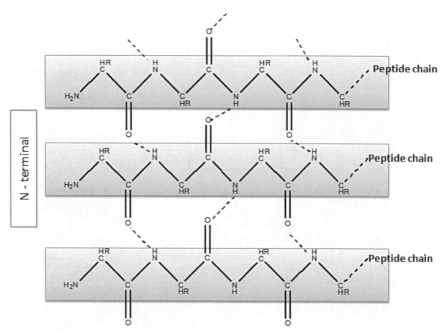

FIGURE 2.26   Schematic representation of parallel β-sheet arrangement (hydrogen bonds between the polypeptide chains [in orange boxes] are indicated by dotted lines).

2.17.2.4   β-STRANDS

The β-strands are small segments of oligopeptides which are present close to each other and form H-bonds. They result in protuberances, coils, and bends in the protein molecule [58].

2.17.2.5   β-TURNS

They are also referred to as hairpin bends. The β turns are essential for folding or turning of the peptide chain and compacting the molecule [59]. The β turns comprise of four amino acid residues. The hydrogen bond in β turn exists between the first and the fourth amino acid residues [60]. Proline residues are found in the β turns. The proline in *cis* configuration results in the sharp β turns in the protein molecule [36]. Glycine is found frequently in the β turns, rendering the flexibility to the turns [1].

2.17.2.6   Ω LOOP

They are commonly found in the globular protein at the surface of the molecule. The Ω loops show multiple H-bonds but no definite or common pattern of bonding. The Ω loops are made of six or more amino acid residues, providing a loop or lariat shape to the structure. They are responsible for tyrosine sulfonation, prohormone cleavage reaction, substrate specificity, protein stability, and protein folding. Moreover, these loops act as screen over the active sites of enzymes. This was first discovered in the enzyme triose phosphate isomerase [61].

2.17.2.7   COLLAGEN

Collagen, an important protein of connective tissue, occurs in skin, tendons, bones, teeth, cartilage, etc. The structure of collagen molecule was given by Ramachandran and Kartha in 1954. The collagen has a triple helix structure, also referred to as tropocollagen. It consists of 3 polypeptide chains, α chain, and more than 1000 amino acid residues long with a repeating unit of Gly-X-Y. The amino acid proline or hydroxyproline are generally present at Y and various other amino acids can occupy position X. Each polypeptide chain of the triple helix is a left-handed helix and all the three polypeptide chains are held together in a right helix. Each polypeptide helix has 10 amino acid residues per three turns and the collagen triple helix has 30 residues per turn, pitch of about 85.8Å, rise of 2.86Å per amino acid residue and twist of −108°. The three helices of the collagen are held together by hydrogen bond between glycine residues and H-bonds formed by water molecules. The structure of tropocollagen and collagen are shown in Figure 2.27.

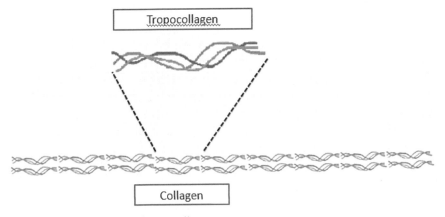

FIGURE 2.27   Collagen and tropocollagen.

There are around 27 types of collagen with 42 different polypeptide chains. In certain collagens, all the three helices are similar but in other types of collagens all the two or three chains may be different designated by αl, α2, and α3 (the difference arises because of various amino acids at X and Y in the repeating unit).

The triple helix or tropocollagen are arranged to form twines that are high in tensile strength and flexible in nature. They are also cross-linked in the connective tissue reducing strain on the tissue. The defects in the collagen structure can result in conditions like Ehlers Danlos syndrome, Osteogenesis imperfect, certain types of osteoporosis, etc. These conditions generally occur because of mutations (single base substitutions) of codon for glycine with some other larger amino acid residue. Mutations leading to changes positions of X and Y either have weak effects [62].

2.17.3   TERTIARY STRUCTURE OF PROTEINS

The tertiary structure of protein is the ultimate compaction of a large poly-peptide or protein molecule [63]. The tertiary structure of protein involves further folding up of the secondary structure and formation of bonds like disulfide bonds, H-bonds, hydrophobic interactions, and salt bridges [1]. This folding of a large polypeptide results in the formation of domains. A domain is a part of protein which performs a specific function [35]. Some examples of proteins with their three-dimensional structure are described below.

2.17.3.1   CYTOCHROME C

Cytochrome C is an important protein of electron transport chain. It contains an iron ion within a heme group which helps in accepting and releasing electron during electron transport chain. This protein plays a key role in ATP synthesis and hence its structure is highly conserved [64]. Human cytochrome C is 104 amino acid residues long with an acylated glycyl residue at N-terminal. The heme unit is attached to the protein by thioether bonds at cysteine residues at positions 14 and 17 [65]. The molecule forms α-helices in the positions 4–14, 51–55, 62–70, 72–75, and 89–102; β-turns at 15–15 and 36–38 and β-strand in the positions 23–25 and 28–30 [66].

2.17.3.2 MYOGLOBIN

The myoglobin is a heme protein of muscles which stores oxygen. It is the first protein to have its three-dimensional structure elucidated by John Kendrew and his colleagues. It is polypeptide of 154 amino acids. The molecule of myoglobin consists of eight α helices, each of which is designated by a letter from A-H. These helices envelop the heme group (contains a porphyrin ring and an iron ion) located between His64 and His93. The iron ion in the myoglobin molecule is bound to six ligands: four nitrogen atoms of pyrroles, imidazole side chain of His93 and the sixth position is occupied by the oxygen molecule. In the deoxymyoglobin, the His93 stabilizes the heme group and shifts iron away from the heme group out of the plane.

Myoglobin stores oxygen in muscles. It helps in controlled release of oxygen during apnea in marine birds and mammals. The myoglobin levels are also elevated in people staying at high altitudes. Myoglobin also serves to buffer oxygen concentrations during increase muscle activity and facilitates oxygen diffusion [67].

2.17.4 QUATERNARY STRUCTURE OF PROTEINS

The quaternary structure of proteins is exhibited those molecules which comprise of more than one polypeptide chains. These polypeptide chains are referred to as subunits. The subunits may be similar or dissimilar [68]. For example, the hemoglobin molecule consists of two α and two β chains. The various subunits of the protein molecule in the quaternary structure are held together by disulfide bonds, H-bonds, hydrophobic interactions, and salt bridges [1]. The quaternary structure of hemoglobin molecule is described below.

2.17.4.1 HEMOGLOBIN

Hemoglobin is a heme protein involved in the transport of oxygen. It comprises of iron ion, porphyrin ring (heme complex), and globin (protein). The hemoglobin molecule is tetramer of two α and two β subunits. The α-subunit is 141 amino acids in length while the β subunit is 146 amino acids in length. The α-subunit has Val-Leu sequence at N-terminal and β subunit has Val-His-Leu sequence at N-terminal [69]. The α-subunit possesses seven α-helices (designated as A-G) and the β subunit comprises eight α-helices (designated as A-H). The subunits are compacted to form a globe-shaped structure with space within each subunit for the heme complex. The heme complex (iron-protoporphyrin IX) consists of porphyrin ring and iron ion, Fe^{2+} as shown in Figure 2.28.

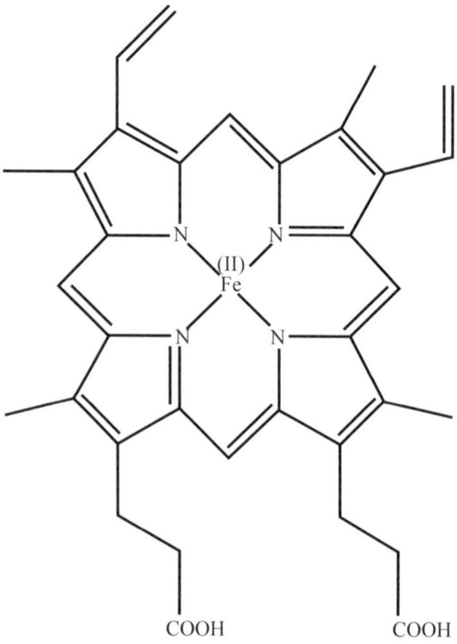

FIGURE 2.28   Heme complex: Iron-protoporphyrin (IX).

The Fe^{2+} is linked by coordinated bonds with four nitrogen atoms of pyrrole ring in a plane, imidazole nitrogen atom of histidine in the F-helix at position 8 (F8) above the plane and oxygen atom below the plane [70]. Phenylalanine of the polypeptide chain aids in positioning of the heme complex within each subunit [71]. The hemoglobin tetramer is shown in Figure 2.29.

The hemoglobin in embryo, fetus, and adult is different. The embryonic hemoglobin is made of two ζ subunits and two ε subunits. The genes responsible for the hemoglobin are downregulated, followed by expression of α and γ genes of hemoglobin, producing fetal hemoglobin (HbF). HbF is the primary hemoglobin during the last two trimesters of pregnancy. HbF has a higher affinity for the oxygen than adult hemoglobin (HbA). The HbF is completely replaced by HbA by the time the baby is one year old. This change initiated at the time of birth with the downregulation of γ gene for hemoglobin and up-regulation of β gene for hemoglobin. The

HbA has another lesser variant HbA2 which consists of two α and two δ subunits, which accounts for 2% of HbA. The HbA accounts for nearly 97% of HbA [70].

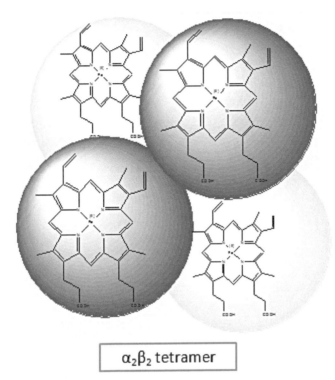

$\alpha_2\beta_2$ tetramer

FIGURE 2.29   Tetrameric hemoglobin A with iron-protoporphyrin (IX) showing 2 α (green) and 2 β subunits.

Oxygen binds with hemoglobin reversibly to form oxyhemoglobin as shown in Figure 2.30 [72].

FIGURE 2.30   Formation of oxyhemoglobin.

There cooperative interaction between oxygen binding sites of hemoglobin, i.e., binding of an oxygen atom eases the binding of other oxygen
atoms to the hemoglobin. This interaction is well understood from the
sigmoid shape of oxygen equilibrium curve. A single hemoglobin molecule
can bind to four oxygen atoms. The oxygenated and deoxygenated hemoglobin is shown in Figure 2.31.

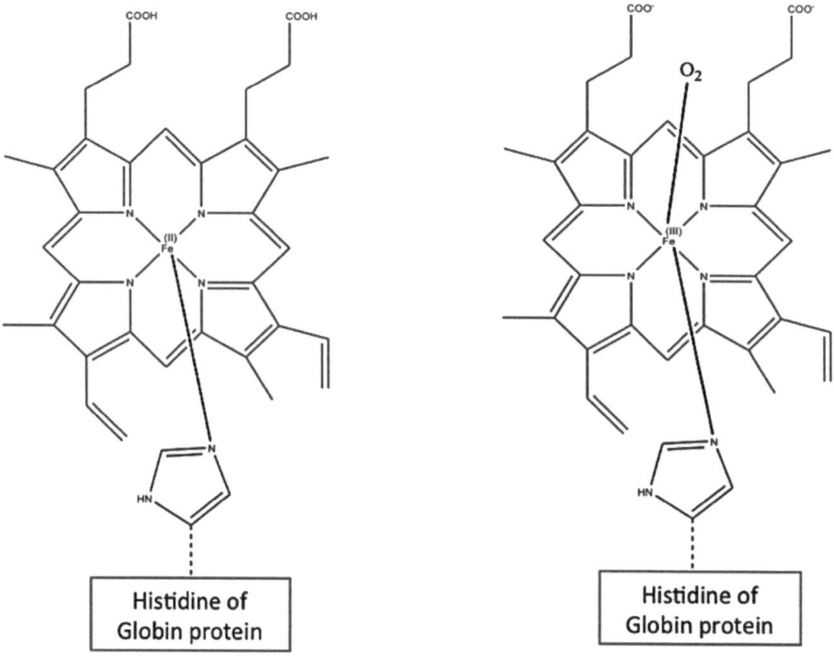

FIGURE 2.31   Oxygenated and deoxygenated hemoglobin.

Hemoglobin is not only involved in transport of oxygen but also helps
in transport of gases like carbon dioxide (CO_2), carbon monoxide (CO), and
nitric oxide (NO) [70]. Carbon monoxide binds more strongly to hemoglobin
and forms carboxyhemoglobin which does not dissociate readily and is the
source of carbon monoxide poisoning [71]. Nitric oxide combines with
deoxygenated hemoglobin to form methemoglobin wherein the iron ion has
+3 charge, Fe^{3+}, and releases nitrate ions. Hemoglobin is regenerated back
from methemoglobin by various mechanisms. Nearly 1% of hemoglobin
is converted to methemoglobin on daily basis. Serious conditions like
methemoglobinemia results on ingestion nitorbenzenes and nitrites, enzyme
deficiencies of as methemoglobin reductase or diaphorases, and abnormal

hemoglobins (HbM). In such conditions, nearly $1/4^{th}$ of hemoglobin becomes non-functional [70, 71]. Mild methemoglobinemia does not require treatment. Ascorbic acid and methylene can be used to treat the conditions [73].

2.17.4.2 SICKLE CELL ANEMIA

Sickle cell anemia is characterized by the sickle-shaped erythrocytes instead of normal biconcave disc-shaped erythrocytes. The hemoglobin of such sickle-shaped erythrocytes is designated as HbS. This transformation occurs because of substitution of valine instead of glutamic acid at position-6 on β-chain of hemoglobin (nucleotide A is replaced by T in the DNA of chromosome 11) [70]. The valine on β subunit orients in the core hydrophobic region of adjacent subunit resulting in an insoluble polymer, damaging the plasma membrane of the erythrocytes. Sickled erythrocytes can aggregate forming microvascular obstructions, besides resulting in conditions like microthrombosis, microembolization, hematocrits, chronic reticulocytosis, etc. Sickle cell anemia is a genetic disorder and only blood transfusion can relieve the severe symptoms of sickle cell anemia. Blood transfusion increases the concentration of normal erythrocytes in the body and helps in better transport of oxygen than the sickle cells [71].

2.18 CLASSIFICATION OF PROTEINS

Around 20 amino acids form numerous proteins. These proteins perform necessary functions of transport, signaling, act as hormones and enzymes; provide immunity (immunoglobulins), etc. The proteins are therefore classified in different ways as follows:

- Based on shape of the protein;
- Based on composition and solubility of the protein;
- Based on function of the protein.

2.18.1 CLASSIFICATION OF PROTEINS BASED ON THE SHAPE OF THE MOLECULE

The proteins can be classified into two groups based on the shape of the molecule:

1. **Globular Proteins:** These have a spherical shape. All the amino acid residues in a globular protein are not in α-helix or β-sheet arrangement. Hydrogen bonding within the polypeptide backbone alone is not sufficient to provide forces for compacting of the molecule. The compact globular shape arises mainly due to hydrophobic interactions amongst the amino acid residues of the protein [74]. The globular proteins exhibit the tertiary as well as quaternary organization of protein. They are soluble aqueous media and are highly diffusible in nature. For example, hemoglobin, cytochrome C, etc. [1].

2. **Fibrous Proteins:** These are thread-like filaments in shape. They are insoluble in aqueous medium [1]. Fibrous proteins possess high tensile strength and can withstand recurring pressure or strain without distortion. Fibrous proteins are chief constituents of extracellular matrix (ECM). Some examples of fibrous proteins are described below:

    i.   **Collagen:** It is a fibrous protein which constitutes nearly 30% of protein in the human body. It provides structure as well as tensile strength to the tissues and organs. Collagens also perform various functions like tissue repair, molecular filtration, cell adhesion, etc.

    ii. **Elastin:** This protein, as the name, has a property of stretching and returning back to the normal shape on the removal of stress. Elastin occurs in vertebrates and is majorly found in the walls of blood vessels in regulation of blood flow. The elastin is arranged in the human body during fetal development. Once positioned, these fibers tolerate many stretching and relaxing cycles all throughout the life of the organism.

    iii. **Fibrillin:** They are extracellular glycoproteins and form microfibils. The protein fibrillin has three isoforms: Fibrillin-1, Fibrillin-2, and Fibrillin-3. Fibrillin-1 is expressed during development of fetus as well as all throughout the life. Fibrillins as a part of microfibril is involved in providing framework to elastin fibers. Fibrillins are also involved in regulating the function of TGF-β family of growth factors [75].

2.18.2   CLASSIFICATION OF PROTEINS BASED ON THE COMPOSITION AND SOLUBILITY OF PROTEINS

It is most widely accepted classification of proteins and has been proposed by the committees of British Physiological society and American Physiological Society [1]. The classification is described as follows:

1. **Simple Proteins:** These are composed of only amino acids. They are classified as follows:

    i. **Protamines:** These are proteins rich in arginine found in nuclei of sperm cells [76]. They are soluble in water and ammonium hydoxide [1]. Protamines are also rich in other basic amino acids [77]. For example, salmine form sperm of salmon, cyprinine from sperm of herring, etc. [1].

    ii. **Histones:** These are basic proteins rich in amino acids lysine and arginine [79]. They are insoluble in ammonium hydroxide. For example, nucleohistones in nucleus [1].

    iii. **Albumins:** They are commonly occurring proteins soluble in water, dilute acids, bases, and salts. For example, serum albumin of plasma and myosin of muscles.

    iv. **Globulins:** They are of two types: Pseudoglobulins and Euglobulins. Pseudoglobulins are soluble in water while euglobulins are not. For example, pseudoglobulin of milk whey and serum globulin (euglobulin) of plasma [1].

    v. **Glutelins:** These are proteins soluble in highly dilute acids and bases but insoluble in neutral solvents. For example, glutenin from wheat, glutelin from rye, and oryzenin in rice [79].

    vi. **Prolamines:** These are proteins soluble in dilute acids and bases and 60–80% alcohol. These proteins are insoluble in water and dilute salt solutions. For example, zein in corn and gliadin in wheat.

    vii. **Scleroproteins:** These are animal proteins insoluble in water, dilute acids, bases, and salts, and 60–80% alcohols [1]. For example, elastin and collagen of extracellular matrix [75].

2. **Conjugated Proteins:** These are simple proteins found in association with molecules other than proteins [80]. They are further classified as:

    i. **Metalloproteins:** These are proteins containing one or more metal ions bound to polypeptide chain at a particular site. The metal ions generally bind at nitrogen atoms of histidine, oxygen atoms of aspartate or glutamate, and sulfur atoms of cysteine reisdues. The metal ions generally found occurring in metalloproteins are calcium, selenium, iron, zinc, copper, etc. Metal ions of these proteins either help in stabilizing the protein molecule or are components of the active sites [81]. For example, iron in siderophilins, copper in ceruloplasmin [1], magnesium in cardiotoxin II, etc. [82].

ii. **Chromoproteins:** These are proteins conjugated with pigment molecules [1]. For example, Hemoglobin, myoglobin, cytochrome C, etc. [83].

iii. **Glycoproteins:** These are sugars conjugated with sugars like D-galactose, D-manose, D-glucose, L-fucose, D-xylose, N-acetyl-D-glucosamine, N-acetyl-D-galactosamine, and various sialic acids [84]. Glycoproteins contain less than 4% of carbohydrates. For example, egg albumin.

iv. **Mucoproteins:** These are proteins conjugated with more than 4% of carbohydrates. For example, ovomucoid of egg white and mucin from saliva [1].

v. **Phosphoproteins:** These are proteins conjugated to phosphoric acid [1]. The phosphorylation of proteins generally occurs at Ser, Thr, or Tyr residues which help in regulating the activity of enzymes, localization in subcellular compartments, protein degradation, and formation of complex [85]. For example, Tyrosine hydroxylase, opioid receptors, ligand gated channels, G proteins, actin, etc. [86].

vi. **Lipoproteins:** These are proteins conjugated with lipid molecules. Lipoproteins help in transport of cholesterol and triacyl glycerides in the plasma. The lipoproteins are broadly classified on the basis of density as: chylomicrons, very-low-density lipoproteins (VLDL), intermediate-density lipoproteins (IDL), low-density lipoproteins (LDL), and high-density lipoproteins (HDL) [87].

vii. **Nucleoproteins:** These are proteins found associated with nucleic acids. For example, histones in the nucleus [78].

3. **Derived Proteins:** These are obtained from simple or conjugated proteins by treating with various physical or chemical agents [80]. The derived proteins are classified as follows:

   i. **Primary Derived Proteins:** These are those proteins whose size does not change much even on treatment with physical or chemical agent. They are further classified as:

      a. **Proteans:** When enzymes, acids or water act on protein, the first produced protein product is referred to as proteans. Proteans are insoluble in water. For example, edastan, and myosin.

    b. **Metaproteins:** These are produced by the action of acid or alkalis at 30–60°C on the proteins. Metaproteins are soluble in dilute acids and bases but insoluble in water. For example, acid metaproteins.

    c. **Coagulated Proteins:** When proteins are heated or acted upon by alcohol, the resulting product is a coagulated protein. For example, egg white on boiling becomes coagulated.

  ii. **Secondary Derived Proteins:** Proteins on hydrolysis produce secondary derived proteins. The resulting molecules are smaller than parent protein molecule [1]. They are further classified as:

    a. **Proteoses:** These are the molecules produced by the action of water or acidic enzymes like pepsin on the proteins. For example, albumoses [88].

    b. **Peptones:** These on further hydrolysis produce peptones. Peptones are soluble in water.

    c. **Polypeptides:** Peptones on further hydrolysis produce polypeptides [1].

2.18.3 *CLASSIFICATION OF PROTEINS BASED ON THEIR BIOLOGICAL FUNCTIONS*

Proteins perform many functions in living organisms. Based on the nature of functions performed by proteins they are classified as follows:

1. Immunoglobulins or Antibodies are proteins which provide immunity in an organism. Antibodies bind to specific antigens present on the surface of infecting bacteria and target them for further destruction by other components of immune system. For example, immunoglobulin G, immunoglobulin M, etc.

2. Enzymes are biological catalysts of cells. Enzymes combine with substrate molecules and modify them to form product molecules. For example, nuclease hydrolyzes nucleotides, phosphatases remove phosphate group form the substrate molecule, etc. [89].

3. Structural proteins like collagen form an important constituent of connective tissue [75].

4. Contractile proteins like tubulin and actin provide structure to the cell. They also help in various functions like locomotion, cell division, and contraction of muscles [90].

5. Transport proteins aid in transport of specific molecules. For example, hemoglobin helps in transport of oxygen and carbon dioxide; lipo-proteins transport cholesterol, and triacylglycerides, etc.
6. Hormonal proteins function to regulate various processes in the body. For example, insulin regulates glucose concentration in blood by increasing its absorption by the cells; oxytocin stimulates contraction of muscles of uterus during parturition, etc.
7. Storage Proteins store and supply necessary nutrients for developing embryo. For example, ovalbumin of eggs, etc.
8. Toxic proteins are found in snake venom which degrades enzymes in the body of the prey [1].

2.19   STRUCTURE-FUNCTION RELATIONSHIP OF PROTEINS

Proteins perform many functions in the body of an organism. The functions of proteins are related to its structure. The structure of a protein is organized into various levels as follows:

- Primary structure;
- Secondary structure;
- Tertiary structure;
- Quaternary structure.

A protein molecule consists of many domains. These domains function as site for substrate binding, membrane anchoring, allosteric site, inhibitory site, etc. For example, protein kinases possess two domains: amino terminal domain binds ATP and carboxyl terminal domain binds the peptide or protein. The transfer of phosphoryl is brought about by the amino acid residues positioned in the loop between the two domains [35]. Thus, the structure of the protein molecule imparts a specific function to it. Structure function relationship of proteins are elucidated below with few examples.

Sickle cell anemia is a genetic disorder resulting from mutation which results in replacement of glutamic acid at position 6 of β chain with valine [91]. The valine residue, being aliphatic in nature, orients in the core hydrophobic region of adjacent subunit. As a result, an insoluble polymer is formed with reduced capacity to carry oxygen. The polymer so formed also damages the plasma membrane of the erythrocytes [71].

Insulin is an important hormone regulating glucose concentration of blood by signaling the cells for its uptake [1]. The signal is mediated at receptor by a dimeric ligand formed by the insulin. This dimeric ligand is formed by the amino acid residues of position 24–26 of B chain. The B24 residue is phenylalanine and B26 residue is tyrosine. These residues form part of hydrophobic core as well as are part of the antiparallel β sheet structure. Any change in these amino acid residues prevents dimer formation and hence, the signal transduction to the cell. Similarly, the amino acid residue at position 23 of B chain is essential for the function of insulin [92].

The lock and key hypothesis of enzyme function given by Fischer also advocates the structure function relationship. The process of denaturation results in complete distortion of protein structure and its activity [93].

A single or a small change in the amino acid sequence of a protein can change its structure and alter its function. Certain diseases/disorders which occur due to change in protein structure are listed in Table 2.8.

TABLE 2.8   Diseases/Disorders with Altered Structure Function Relationship of Protein

Name of the Disorder/Disease	Name of the Defective Protein
Alzheimer	β-Amyloid
β-Thalassemia	Hemoglobin
Creutzfeldt-Jakob disease	Prion
Sickle Cell Anemia	Hemoglobin

Thus, the structure of protein is essence to its function. The crucial structure function relationship can be analyzed by various tools like X-ray crystallography, NMR spectroscopy, mutation studies, and bioinformatics tools as shown in Table 2.9 [93–95].

TABLE 2.9   Tools/Techniques Used for Determining Structure Function Relationship of Proteins

Tools/Techniques	Functions
BLAST	Prediction of protein function
FASTA	Prediction of protein function
Epitope tagging	Locating the protein within the cell
Affinity chromatography	Detecting the protein-protein association
Immunoprecipitation	Detecting the protein-protein association
Phage display technique	Detecting protein-protein interaction
Surface plasmon resonance (SPR)	Detecting protein-protein interaction
DNA footprinting	Locating the sites of protein binding on DNA

2.20   PROTEIN STRUCTURE DETERMINATION

In order to understand the function of protein it is imperative that it's three-dimensional structure is determined. The three-dimensional structure of protein is determined using various techniques described below:

1.   **X-Ray Crystallography:** Myoglobin was the first protein whose structure was determined by using John Kendrew in 1960 by x-ray crystallography. This was followed by determination of many other structures of protein using the approach of x-ray crystallography [35]. X-ray crystallography helps locate the exact position of atom in the protein molecule. This method is one of the apt methods for protein structure determination as the x-rays used in the method have wavelength same as that of the lengths of covalent bond [96].

The x-rays have a wavelength of about 1Å. The x-rays are allowed to hit the protein crystal and the x-ray diffraction pattern is recorded. Most of the rays pass through the crystal without any diffraction but only a few rays are diffracted [97]. The diffracted waves generated when the x-rays hit the protein crystal, are dependent on the following points:

i.   The amplitude of scattering of rays is directly proportional to number of electrons in the atom.
ii.   If the diffracted waves are in phase they reinforce each other.
iii. If the diffracted waves are out of phase they cancel each other.
iv.   The atomic arrangement in the molecule determines the diffraction pattern of rays [96].

The principle of X-ray crystallography is illustrated in Figure 2.32.
The major steps in the process of x-ray crystallography are described below:

• **Step 1:** Preparation of protein crystals using various conditions like pH, temperature, etc., and various reagents like salt solutions, organic solutes, etc.
• **Step 2:** Protein crystals are mounted on the quartz capillaries and exposed to x-rays of wavelength of 1.5 Å to confirm whether the crystals are of protein or salt. After the confirmation, crystals are frozen in liquid nitrogen for further analysis [35].

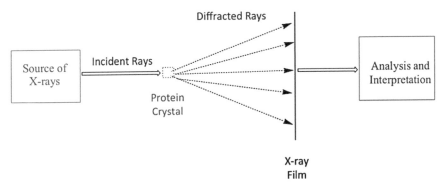

FIGURE 2.32 Principle of x-ray crystallography.

- **Step 3:** The protein crystals are rotated and exposed to x-rays and diffraction pattern is detected on x-ray film or solid state electronic detector.
- **Step 4:** The diffraction pattern of waves is analyzed using Fourier transform/synthesis. The phase of the waves must be determined to understand intensity of the spot and hence the amplitude of waves. Isomorphous displacement is done to solve this problem using mercury or uranium which binds cysteine residues or selenium in the form of selenomethionine within the protein crystal [35]. These atoms behave reference marker within the protein crystals.
- **Step 5:** An electron density map or sketch is obtained which depicts the position of atoms in the protein crystal [96].

The stepwise representation of the process is given in Figure 2.33.

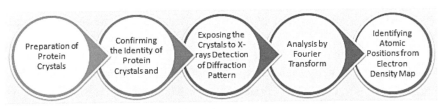

FIGURE 2.33 Steps of x-ray crystallography.

The crystals of some protein retain their function even in the crystallized form and are true representative of protein molecule in x-ray crystallography procedure. The crystallography of such crystals makes use of Laue approach which does not require rotating of the crystal and long exposure duration for x-rays [35].

2. **NMR Spectroscopy:** It refers to the Nuclear Magnetic Resonance Spectroscopy. This technique is used to analyze the structure of small protein molecules in solutions. The NMR spectroscopy has certain advantages as given below:

    i.   It does not require use of crystallized molecule of protein as in x-ray crystallography.

    ii. It requires only little volume of sample solution for spectroscopic determination [97].

    iii. It allows analysis of changes in protein conformation on binding of ligand or substrate [35].

The NMR spectroscopy is based on the fact that protons or hydrogens possess magnetic spin. In the presence of magnetic field these protons orient in such a way that their spin align with the direction of magnetic field. When such protons are exposed to radiofrequency, they become excited due to absorption of energy. The protons return to the ground state by emitting radiation of radio frequency in the form of spectra. The spectra shown by a hydrogen atom or proton is characteristic of the environment in which it is placed, i.e., the nucleus of one atom absorbing radiation effects the absorption and emission of radiation by its neighboring atomic nuclei [97]. This change in absorption or emission is observed due to chemical shift. The electrons present in an atom around the nucleus also produce a small magnetic field which opposes the applied magnetic field producing a shielding effect, thus affecting the amount of radiation absorbed. The chemical shift of the molecule of interest is measured using trimethyl silane as a reference molecule.

The protons therefore behave characteristic to the environment in which they are present, with each group on the molecule behaving in a highly specific manner. The NMR spectroscopy is very useful for small proteins ≤ 40 kDa in size [96]. Thus, NMR spectroscopy aids in determining the structure of proteins in solution as well as molecules like RNA, protein side chains of glycoproteins, etc. [97].

2.21 PHYSICAL PROPERTIES OF PROTEINS

Proteins exhibit various physical properties described as follows:

- **Color:** Proteins are generally colorless but chromoproteins like hemoglobin are red in color due to presence of pigment heme.
- **Taste:** Proteins are generally tasteless [1].
- **Shape:** Proteins may be globular or fibrillar in shape. For example, Hemoglobin, cytochrome C, etc., are globular proteins; Collagen, fibrin, elastin, keratin, etc., are fibrillar proteins [98].
- **Size and Molecular weight:** The proteins may be of different sizes and molecular weights as shown in Table 2.10.

TABLE 2.10 Different Proteins with Their Size and Molecular Weight

Name of the Protein	Size	Molecular weight (kDa)
Insulin	50Å (diameter)	5,808
Thrombin	30Å (diameter)	37,000
Ribonuclease A	$38 \times 28 \times 28$ Å^3	13,700
Ferritin	122Å (diameter)	450,000

2.21.1 SOLUBILITY

Protein solubility in a solvent is determined by intrinsic factors (protein-protein, protein-water, protein-ion, and ion-water interactions) and extrinsic factors (ionic strength, solution composition, pH, and temperature). Protein solubility is important for proteins in pharmaceutics. Proteins with low solubility are those proteins which are soluble in relative concentrations of 1–100 mg/ml. Solubility of protein can be increased by adding additives like amino acids like arginine, glutamine, etc., and site directed mutagenesis [99].

The proteins are least soluble at its isoelectric point as it is neutral in nature. Solubility of the proteins increase with increased acidic or basic nature of the solution as at these pH conditions proteins exist as positively charged cations or negatively charged anions [1].

Proteins that are highly soluble can be precipitated by salting-out technique which makes use of salts like ammonium sulfate [99]. The solubility of proteins like globulins in water can be increased by adding salts like sodium chloride [98]. Thus, solubility of proteins can be altered using neutral salts.

2.21.2 DENATURATION

The changes in protein structure due to disruption of disulfide bonds, H-bonds, hydrophobic interactions, salt bridges, etc., leads to loss of biological function. This process is referred to as denaturation. The peptide bonds are not destroyed during the process of denaturation [100]. Denaturation occurs because of various physical or chemical agents. The physical agents causing denaturation are [1]:

- agitation;
- heating;
- cooling and freezing;
- exposure to radiation like UV rays, x-rays, etc.;
- ultrasonic waves;
- high pressure.
- chemical agents causing denaturation are:
- acids and bases [100];
- organic solvents [1];
- detergents;
- chaotropic agents like urea; and guanidium ion [22].

The addition of acids to the solution of protein, leads to protonation of carboxyl groups of acidic amino acid residues resulting in conformation changes and denaturation. Similarly, addition of base to the solution of protein results in deprotonation of side chains of amino acid residues, changing the protein conformation and hence denaturation [100]. Detergents denature protein by disturbing the hydrophobic interactions within the molecule. Chaotropic agents enhance the solubility of non-polar solvents which interfere with hydrophobic interactions in the protein molecule resulting in denaturation [22]. Denaturation of egg albumin on boiling is shown in Figure 2.34.

Denaturation of certain proteins is reversible and is referred to as renaturation. The renaturation was first elucidated in ribonuclease A (RNase A) by Christian Afinsen in 1957. RNase A is denatured in presence of β mercaptoethanol in 8M urea which breaks the disulfide bonds in the protein. The protein is renatured back by dialysis and subsequent exposure of the protein solution to oxygen the renatured protein was enzymatically active as well as similar to its native conformation. The work of afinsen confirmed that the three-dimensional structure of protein is dependent on its primary structure [22]. Similarly, trypsin is denatured at temperatures of 80–90°C and renatured when cooled to 37°C [1].

FIGURE 2.34   Denaturation of egg albumin. (a) Egg white-albumin; (b) egg white-albumin denatured.

2.21.3   CURDLING OF MILK

Souring of milk is a result of bacterial action. The bacteria present in milk convert the sugars to lactic acid. As the lactic acid accumulates, the pH decreases resulting in denaturation of milk proteins, resulting in curdling. Similarly, boiling of egg results in denaturation of albumin protein [100].

2.22   CHEMICAL PROPERTIES OF AMINO ACIDS AND PROTEINS

The chemical properties of proteins can be understood by the following classification:

- Hydrolysis of proteins;
- Reactions with $-NH_2$ group;
- Reactions with $-COOH$ group;
- Reactions with both $-NH_2$ and $-COOH$ groups;
- Reactions with R groups.

The various reactions of proteins are described in subsections.

2.22.1   HYDROLYSIS OF PROTEINS

The first step of amino acid analysis of proteins necessitates the need for hydrolysis [102]. The proteins can be hydrolyzed in the following ways:

1.   **Acid Hydrolysis:** Proteins are hydrolyzed using 6N HCl at 110°C for about 22 hours in slow hydrolysis process [101]. The disadvantages of acid hydrolysis are as follows:
     •   Asparagine and glutamine are converted to aspartic acid and glutamic acid respectively.
     •   Loss of tryptophan in the process.
     •   Serine and threonine are incompletely destroyed.
     •   Unfinished hydrolysis of β branched amino acids (Val-Val, etc.) [101].
2.   **Alkaline Hydrolysis:** Proteins are hydrolyzed using 2N NaOH. The various disadvantages of the process are as follows:
     •   Amino acids like serine, threonine, cysteine, etc., are destroyed.
     •   Amino acids lose their optical activity.
3.   **Enzymatic Hydrolysis:** This is a very mild process. It is highly specific and occurs at optimum temperatures. The enzymes which are used for enzymatic hydrolysis are papain, trypsin, pepsin, bromelain, etc. Enzymatic hydrolysis generates peptides of various lengths [103]. For example, trypsin can be used to cleave the peptide at lysine and arginine residues [104]; papain cleaves the peptide chain adjacent to arginine, lysine, and phenylalanine [103].

The hydrolysis of proteins is shown in Figure 2.35.

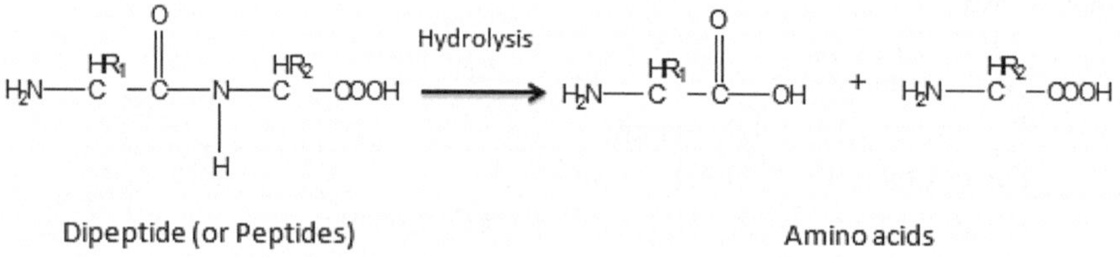

Dipeptide (or Peptides)   Amino acids

FIGURE 2.35   Hydrolysis of proteins.

2.22.2   REACTIONS WITH –NH₂ GROUP

1.   **Acylation:** In the process of acylation, the amino acids are subjected to reaction with acid chlorides or acid anhydrides in presence of

alkaline medium [1]. Acylation helps in protecting the amino group form nucleophilic substitution reaction. Amino acids are protected during peptide synthesis by acylation with benzyl chloroformate by formation of benzyloxycarbonyl derivative [100].

2. **Reaction with Benzaldehyde:** Amino acids react with benzaldehyde to form Schiff's bases [1]. Metal complexes of amino acid Schiff bases possess anti-carcinogenic activity [105].

3. **Reaction with Dansyl Chloride:** Amino group of amino acids react with dansyl chloride (1-dimethylaminonaphthalene-5-sulfonyl chloride) to form dansyl derivatives. These dansyl derivatives resist acid hydrolysis and therefore can be used for establish the N-terminal amino group or ε-amino group of peptides and proteins [106]. Dansyl derivatives are highly fluorescent and can be identified by performing thin layer chromatography on sheets of polyamide. This reaction was originally developed for detection amino acids in peptides but can also be used for proteins [107].

4. **Reaction with Formaldehyde:** Amino acids react with formaldehyde to form a methylol and dimethylol structure. This is a reversible reaction [1, 108, 109]. This reaction forms the basis of formol titration method given by Sorensen. Amino acids cannot undergo direct titration as they exist as zwitterions. Amino acids are first neutralized using an alkali followed by reaction with formaldehyde. The amino acid then behaves as an acid and can be used for regular acid-base titrations [109].

5. **Reaction with Acids:** Amino acids react with acids to form salts. Amino acids are reacted with hydrochloric acid to form chloride salts [1].

6. **Van Slyke Reaction:** Amino acids react with nitrous acid to form α-hydroxy acids with the evolution of nitrogen gas [1]. The amount of nitrogen gas evolved aids in the estimation of aliphatic amino acids [111]. This reaction was used by Van Slyke in 1911 for determination of aliphatic amino acids. This reaction is not given by proline [112].

7. **Oxidation:** The amino acids can be oxidized using potassium permanganate or hydrogen peroxide to form an imino acid [110].

8. **Reaction with Sanger's Reagent:** The Sanger's reagent is 1-fluoror-2, 4-dinitrobenzene (FDNB). FDNB reacts with amino acid to form crystalline yellow-colored N-2,4-dinitrophenyl amino acids [106]. This reaction was used by Fredrick Sanger to determine the sequence of insulin. This reaction also helps in differentiation of lysine present within the peptide and lysine at the N-terminal of the peptide. Sanger was awarded Nobel Prize for his work in 1958 [35].

The reactions of $-NH_2$ group of peptides and amino acids are shown in Figure 2.36.

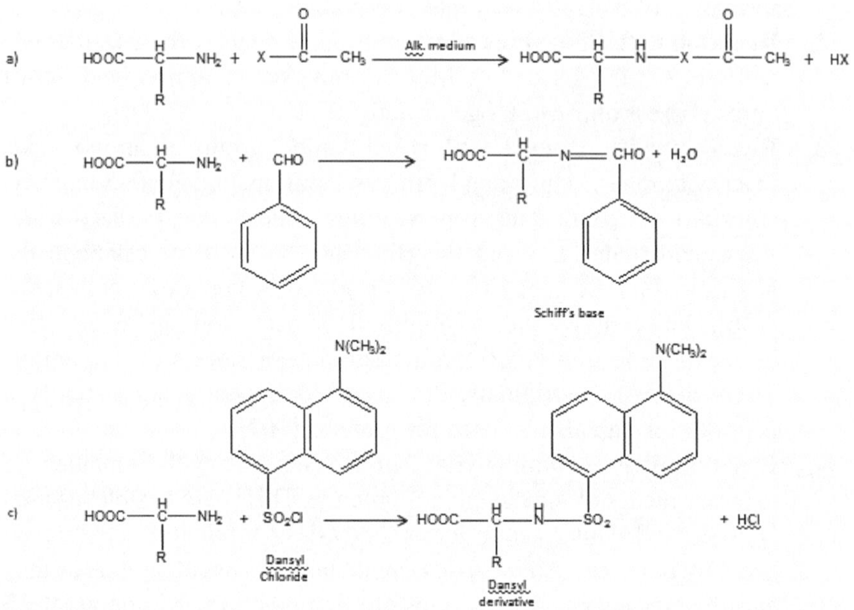

FIGURE 2.36   Reactions with $-NH_2$ group. (a) Acylation; (b) reaction with benzaldehyde; (c) reaction with dansyl chloride.

2.22.3   REACTIONS WITH -COOH GROUP

1. **Formation of Amides:** Amino acids react with amines to form amides [1].
2. **Reaction with Bases:** Amino acids react with bases like sodium hydroxide to form sodium salts of amino acids [1].

    Classic example of sodium salt of glumatic acid Monosodium Glutamate (MSG): MSG is the sodium salt of glutamate. Kikunae Ikeda, a Japanese chemist, first isolated MSG from *Laminaria japonica* and patented the procedure for isolation of MSG from wheat flour. MSG was produced commercially under the name of "Ajinomoto" in 1909. MSG has an umami taste or a meaty taste. MSG also has the tendency to enhance the flavor of the food and is used to enhance the taste of Sake, Japanese rice wine [113].

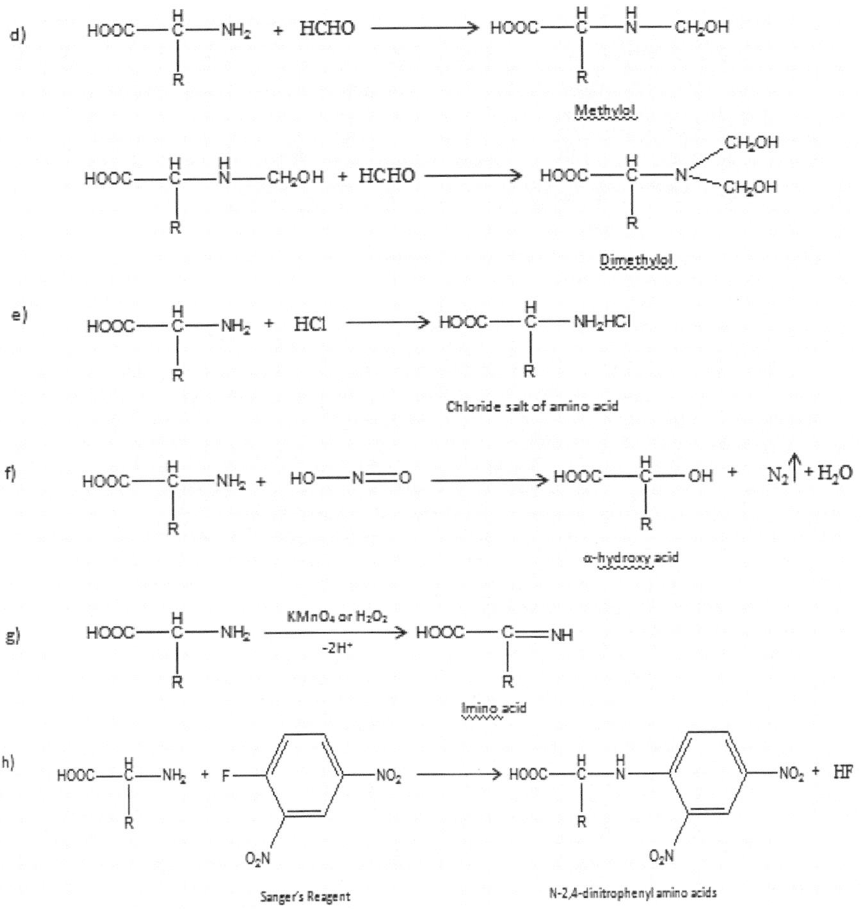

FIGURE 2.36   (d) reaction with formaldehyde; (e) reaction with acids; (f) van slyke reaction; and (g-h) oxidation.

Chinese restaurant syndrome or Kwok's Quease refers to the symptoms experienced by people 20 minutes after consumption of Chinese food [114]. The symptoms like chest pain, flushing, headache, numbness, or burning in or around the mouth, sense of facial pressure or swelling, and sweating are observed [115]. These symptoms are not experienced by every person consuming Chinese food but by those people who are allergic to MSG. The symptoms observed subside after sometime without any further effects [114].

3.   **Decarboxylation:** Amino acid are decarboxylated by heating with $Ba(OH)_2$ or diphenylamine to form an amine [110].

4. **Esterification:** The carboxyl group of amino acids is esterified with alcohol in presence of gaseous HCl [100]. The esters of amino acids were used by Emil Fischer for isolation of different amino acids by fractional distillation [106].
5. **Reduction:** Amino acids are reduced to alcohols by sodium/lithium borohydride and iodine in tetrahydrofuran [112].

The reactions of –COOH group are shown in Figure 2.37.

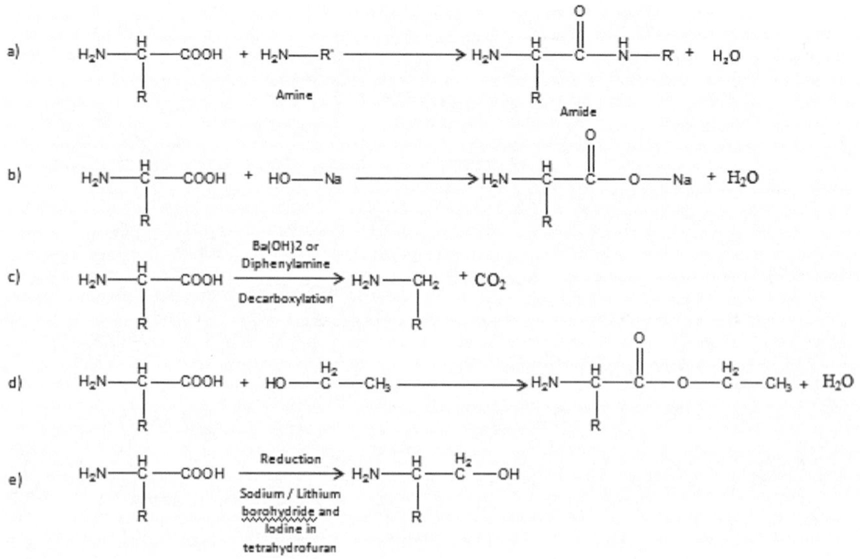

FIGURE 2.37 Reactions with –COOH group. (a) Formation of amides; (b) reaction with bases; (c) decarboxylation; (d) esterification; and (e) reduction.

2.22.4   REACTIONS WITH –NH₂ AND –COOH GROUP

1. **Reaction with Carbon Disulfide:** Amino acids react with carbon disulfide to form 2-thio-5-thiozolidones [1].
2. **Edman Degradation:** Amino acids of peptides react with phenyl isothiocyante to form phenyl thiohydantoic acid of the peptide. This derivative is cleaved using acid in solvent like nitromethane to yield phenylthiohydantoin and a peptide shorter by one amino acid. This method was first used by Pehr Edman for the labeling of N-terminal amino acid of the peptide [35].

3. **Ninhydrin Reaction:** It was discovered by Siegfried Ruhemann in 1910 [116]. Ninhydrin is 1,2,3-indantrione monohydrate. Amino acids react with ninhydrin in the presence of pyridine to form a purple-colored complex called Ruheman's purple. Aldehyde and carbon dioxide are also produced in this reaction [100]. Ninhydrin reaction is used for spectrophotometric estimation of amino acid with absorption maximum = 570 nm. Proline produces yellow-colored compound when subjected to ninhydrin reaction (the absorption maximum = 440 nm) [106].

    *   **Ninhydrin in Forensics:** The ninhydrin reaction forms the basis of fingerprints detection in forensic science. Amino acids are released form dermal ridges of the fingers. When ninhydrin and pyridine is sprayed, the purple-colored fingerprints are observed [100].

4. **Reaction with Phenyl Isocyanate:** Amino acids react with phenyl isocycanate to form hydantoic acid. Hydantoic acid is converted to hydantoin by boiling in an acidic medium [106].

5. **Reaction with Phosgene:** Amino acids react with phosgenes to form N-carboxy anhydrides. N-carboxy anhydrides can be used for synthesis of polyamino acids [117].

The reactions are shown in Figure 2.38.

2.22.5 REACTIONS WITH R GROUPS

1. **Biuret Reaction:** Proteins react with solution of copper sulfate in sodium hydroxide to form a violet-colored complex [118]. The compound biuret (shown in Figure 2.35) also gives this reaction and hence the name. The intensity of the color formed is directly proportional to the number of peptide bonds and thus, the protein concentration, forming a deep blue-violet complex as shown in Figure 2.39. This test is not given by dipeptides [1]. Biuret reaction is also used for quantitative estimation of proteins by reading absorption at 540 nm [119].

2. **Folin's Test:** Tyrosine forms a blue colored complex when it reacts with phosphomolybdotungstic acid in alkaline medium [1].

3. **Pauly Test:** Tyrosine and histidine react with diazotized sulfanilic acid to form a red-colored azo compound [121].

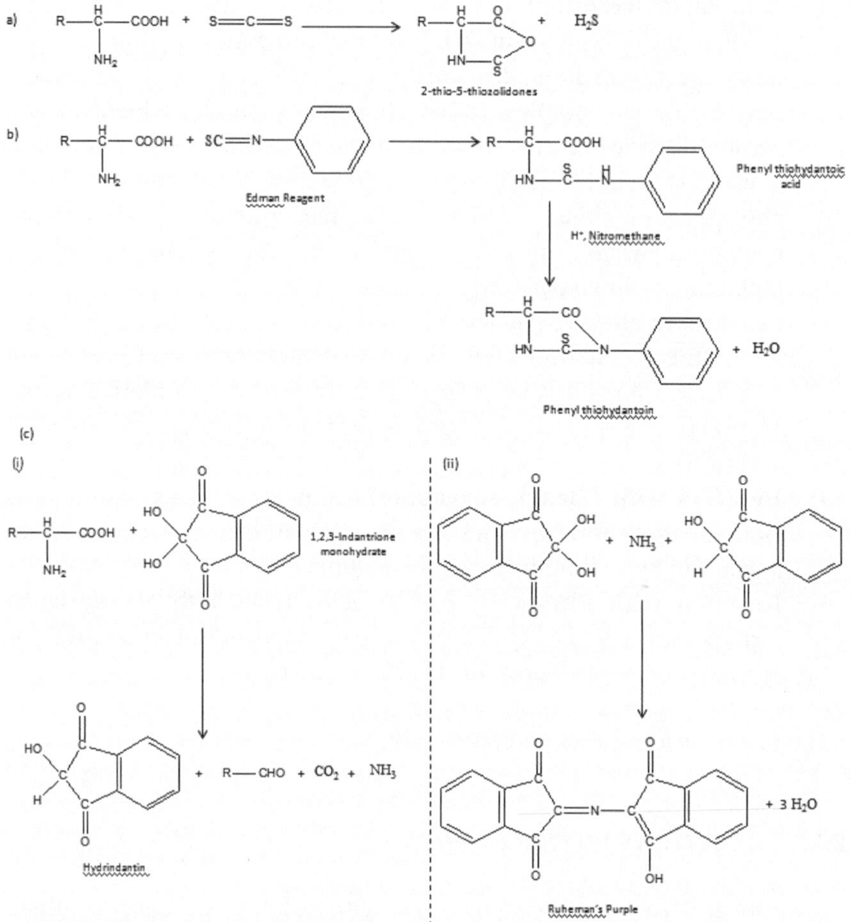

FIGURE 2.38 Reactions with –NH$_2$ and –COOH group. (a) Reaction with Carbon disulfide; (b) Edman degradation; (c) ninhydrin reaction.

4. **Sakaguchi Test:** Arginine reacts with α-naphthol and sodium hypo-bromite to form a red-colored compound [121].

5. **Xanthoprotic Test:** Amino acids like tyrosine, tryptophan, and phenylalnine react with concentrated nitric acid to form yellow-colored nitro compounds. These compounds turn orange on addition of a base [1].

6. **Sullivan Test:** Cysteine reacts with sodium-1, 2-naphthoqui-none-4-sulfonate and sodium hydrosulfite to form a red-colored compound [1].

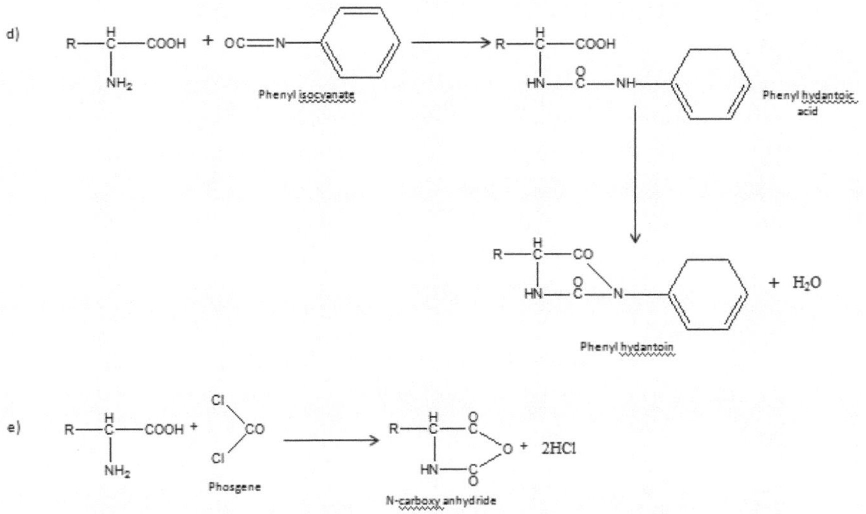

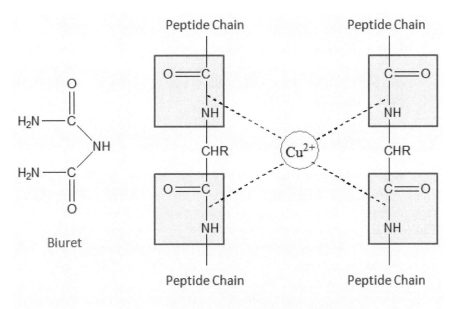

FIGURE 2.38　(d) phenyl isocyanate; and (e) reaction with phosgene.

FIGURE 2.39　Compound biuret and colored complex of biuret reaction.

2.23   CHEMICAL REACTIONS USED FOR QUANTITATIVE ESTIMATION

1.   **Bichinchonic Assay:** The Bichinchoninic acid (BCA) assay is highly sensitive method commonly used for protein estimation. The assay is based on two-step reaction:

   - **Step 1:** Formation of cuprous ions (Cu^{2+}) from cupric ions (Cu^+) by the peptide bonds and certain amino acid residues in alkaline medium.
   - **Step 2:** Chelation between two molecules of BCA and Cu^+ to form a blue colored complex with an absorption maximum at 562 nm [121].

2.   **Bradford Method:** This method of protein estimation is based on the association of amino acid residues arginine, lysine, and histidine with coomassie brilliant blue (CBB) dye in acidic conditions. The complex formed by protein and CBB is blue in color and has an absorption maximum of 595 nm [120].

3.   **Lowry Method:** Proteins react with solution of alkaline copper sulfate and phosphomolybdotungstic acid to form a blue colored complex [120]. The quantification can be done by reading the absorption at 750 nm [119].

2.24   SECRETION OF PEPTIDES

The peptides or proteins are synthesized by translation of mRNA. The mRNA is transcribed in the nucleus from the DNA template. The peptides produced after translation undergoes folding and various modifications to form a functional protein. The schematic representation is given in Figure 2.40.

The peptide secretion is a complex process. A polypeptide undergoes several modifications before it is secreted out of the cell. The process of peptide secretion is described below in detail.

The synthesis of secretory peptides/proteins is initiated in the cytoplasm. The mRNA molecules bound to 80S ribosome begin the synthesis of the secretory peptides/proteins. As the synthesis of protein proceeds, a 16–30 amino acid residue signal sequence, ER signal sequence, at N-terminal guides the ribosome towards the membrane of ER. The signal sequence comprises of the positively charged amino acid residues and 6–12 hydrophobic amino acid residues.

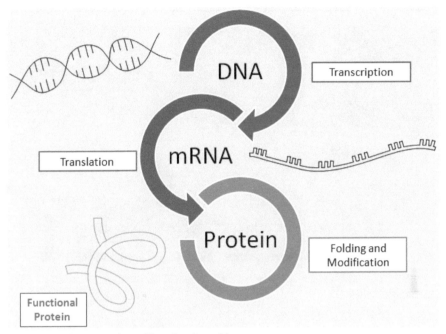

FIGURE 2.40   Formation of functional peptide.

The signal sequence is recognized by signal recognition particle (SRP) present in the cytoplasm. SRP bind to signal sequence of the nascent polypeptide and 60S subunit of the ribosome forming a complex. This complex binds to the SRP receptor present on the membrane of ER. The SRP receptor consists of two subunits: α subunit made of 640 amino acids and β-subunit made of 300 amino acids. The SRP receptor has α-subunit on the surface of the ER membrane while β subunit is present as transmembrane protein. The SRP, nascent polypeptide, and 60S-subunit of ribosome complex bind to SRP receptor at α-subunit.

Hydrolysis of GTP molecule brings about the release of SRP and SRP receptor from the nascent polypeptide, followed by binding of the signal sequence to translocon on the ER membrane. The translocon channel is closed by small part of Sec61p protein. Binding of signal sequence to the translocon opens the channel. The growing polypeptide chain then enters the ER through the translocon channel. The growing polypeptide enters the translocon in the form loop. This loop comprises of signal sequence and nearly 30 adjacent amino acid residues. The signal sequence of the polypeptide is cleaved and subsequently the growing polypeptide enters the lumen of ER [122]. The schematic representation is given in Figure 2.41.

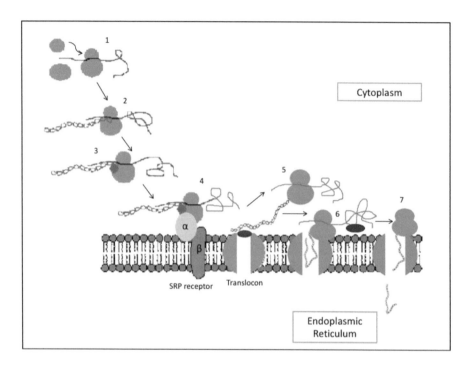

FIGURE 2.41   Process of peptide secretion. (1) mRNA molecules bound to 80S ribosome begin the synthesis of the secretory peptides/proteins. (2) A signal sequence, Endoplasmic reticulum signal sequence (in pink), at N-terminal guides the ribosome towards the membrane of endoplasmic reticulum (ER). (3) The signal sequence is recognized by signal recognition particle (SRP, indicated in red) present in the cytoplasm. SRP binds to signal sequence of the nascent polypeptide and 60S subunit of the ribosome forming a complex. (4) This complex binds to the SRP receptor present on the membrane of ER. The SRP, nascent polypeptide and 60S subunit of ribosome complex binds to SRP receptor at α subunit. Hydrolysis of GTP molecule brings about the release of SRP and SRP receptor from the nascent polypeptide, followed by binding of the signal sequence to translocon on the ER membrane. (5) The translocon channel is closed by small part of Sec61p (in deep black) protein. Binding of signal sequence to the translocon opens the channel. (6) The growing polypeptide chain then enters the ER through the translocon channel. The growing polypeptide enters the translocon in the form loop. (7) The signal sequence of the polypeptide is cleaved and subsequently the growing polypeptide enters the lumen of ER.

The polypeptide is folded by chaperons to attain their three-dimensional conformation. The disulfide bonds are formed in the ER by the protein disulfide isomerase enzyme. N-linked glycosylation of proteins occurs in the ER. Glycosylation occurs at asparagine residues in the sequence Asn-X-Ser/Thr by the enzyme oligosaccharyl transferase. A 14 sugar unit of oligosaccharide is added on to the asparagine residue. After glycosylation a 4 sugar residue fragment (composed of three glucose and one mannose) is removed within the lumen of ER [123]. The proteins are transported to Golgi apparatus for modifications (like glycosylation [125]).

The secretory proteins are transported to cis-Golgi by transport vesicles. The transport vesicles bud off from the surface of ER and fuse with membrane of cis-Golgi releasing the protein within its lumen [123, 124]. The transport vesicles can also fuse to form cis-Golgi network. The cis-Golgi moves from its cis position to trans position by a process called cisternal progression. During this process, a new stack of Golgi gradually moves ahead away from the ER becoming medial and finally trans-Golgi cisternae. The proteins are constantly moved from the advancing cisternae to the former cisternae by transport vesicles. This mode of vesicular transport is referred to as retrograde transport. The secretory proteins reach trans-Golgi network mediated by cisternal progression as shown in Figure 2.42.

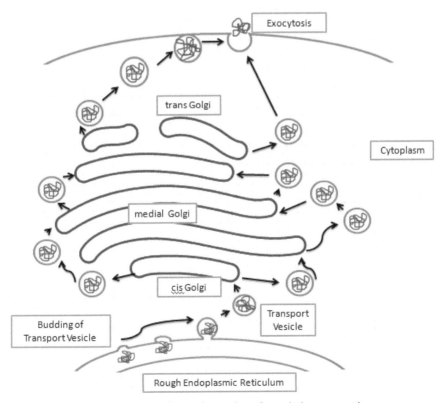

FIGURE 2.42   Cisternal progression and secretion of protein by exocytosis.

There are two types of secretion from the Golgi apparatus based on the requirement of the protein. They are:

i.   Constitutive secretion; and
ii. Regulated secretion.

In the constitutive secretion, the proteins that are continuously secreted are packed in the transport vesicles which instantly move to the plasma membrane and fuse with it, releasing the proteins by exocytosis. For example, albumin secretion by liver hepatocytes; collagen secretion by fibroblasts, etc.

In the regulated secretion, the secretory proteins are loaded into the secretory vesicles and stored inside the cells. The secretory vesicles are exocytosed on receiving an appropriate signal. For example, release of insulin from β islet cells of pancreas; release of casein from mammary glands, etc. [124].

2.25 MODIFICATIONS IN PEPTIDE TOPOLOGY

The structure of protein is governed by primary, secondary, tertiary, and quaternary levels of organization. The structure of protein is closely associated with the function of protein. Besides, the surface of the protein molecule also plays an important role in protein function. The ligand interacts with the surface of receptor. The specificity, affinity, and recognition of receptor are governed by the surface of protein and in turn, the residues present on the surface. Small changes in the amino acid residue of the peptide provide idea about the topology and the conformation.

The modification of peptide backbone may provide insight into the structure-function relation of the peptide/protein. The various topological modifications created in the protein structure involve:

i.   Pseudopeptide (ψ) bonds;
ii. Retro-inverso peptides.

2.25.1 PSEUDOPEPTIDE (Ψ) BONDS

Pseudopeptide bonds are formed by replacement of the peptide bond with isosteric group. The introduction of pseudopeptide bond in the peptide backbone does not change the size of the protein [126]. The first isosteric unit was a methylene thioether bond (CH_2S) (used in 1977) in the glycyl leucine peptide as an analog against aminopeptidase M.

The pseudopeptide bonds in peptides are mentioned according to nomenclature as:

Sequence of the peptide – ψ [isosteric group] – sequence of the peptide

Example: Gly-Phe-Met-Arg-ψ [CH_2S]-Trp-Tyr-Lys

The ψ indicates the psuedopeptide bond and [ ] indicates the type of isosteric group [127]. The examples of pseudopeptide bonds are given in Table 2.11.

TABLE 2.11 Pseudopeptide Bonds and its Nomenclature

Pseudopeptide Bond	Name of the Bond	Nomenclature
–CH_2S-	Thioether	Ψ[CH_2S]
–COO-	Ester	Ψ[COO]
–COS-	Thioester	Ψ[COS]
–$COCH_2$-	Ketomethylene	Ψ[$COCH_2$]
–CH_2NH-	Reduced carbonyl	Ψ[CH_2NH]
–CSNH-	Thioamide	Ψ[CSNH]

The ψ [CH_2NH] is a frequently used pseudopetide bond due to ease of incorporation in the peptide during synthesis and can provide receptor antagonists [126].

2.25.2 RETRO-INVERSO PEPTIDES

The biologically active peptides/proteins have a very short half-life. This short half-life limits the therapeutic use of the peptides. This problem of short half-life can be resolved by using retro-inverso peptides. The retro-inverso peptides are those peptides which possess one or more reversed peptide bonds in the peptide backbone. The reversal of the peptide bond is brought about by using substituted gem-diamino dialkyl residue as N-terminal residue (represented as g Xaa where g is gem-diamino dialkyl and Xaa is a three-letter symbol of amino acid) and substituted malonic acid residue (represented as m Xaa where m is malonic acid and Xaa is a three-letter symbol of amino acid) as C-terminal residue. The major advantage of this bond reversion is retention of planar characteristic of the bond as well the conformation of the bond as that in the original parent molecule. Besides, the retro-inverso peptides have also been found to possess enhanced activity and stability than the original molecule [126].

2.26 PEPTIDES HORMONES OF BIOLOGICAL IMPORTANCE AND PEPTIDE DRUGS

Hormones are important biochemical messengers in the body which help in regulating various functions. The hormones are classified as:

- Steroid hormones;
- Peptide hormones;
- Hormones as amino acid derivatives.

The various hormones which are categorized in the above-mentioned classes and the organs secreting them are listed in Table 2.12.

The peptide hormones and various peptide drugs (agonists and antagonists are shown in Figure 2.43) are described in the subsequent sections.

2.26.1 HORMONES OF HYPOTHALAMUS

Hypothalamus is a small organ of the brain involved in functions like release of hormones from many glands, especially the pituitary gland sex drive, sleep, thirst, heart rate, body temperature, hunger, and mood [126]. The hypothalamus regulates various functions of the body by producing hormones. The hormones produced by the hypothalamus act on the anterior pituitary (adenohypophysis). The anterior pituitary produces hormones called releasing hormones/factors or inhibitory factors under the influence of hypothalamic hormones. The cholinergic and dopaminergic stimuli control the release of hypothalamic hormones. The synthesis of hypothalamic hormones is also regulated by feedback mechanism. The GnRH of hypothalamus and the peptide drugs are described below.

2.26.1.1 GONADOTROPIN RELEASING HORMONE (GONADOLIBERIN, GNRH)

GnRH is a decapeptide (figure) which stimulates the release of luetinizing hormone (LH) and Follicle-stimulating hormone (FSH) from the anterior pituitary. Thus, this hormone influences the fertility in males and females [129]. The GnRH therefore is an important target molecule for chemist developing drugs for fertility and anti-fertility issues [126].

TABLE 2.12   Types of Hormones and Organs Producing Them

Type of Hormone	Name of the Organ	Hormone Produced
Steroid hormones	Ovary	β-estradiol, Estriol, Estrone
	Testes	Testosterone, Androsterone, Dehydroepiandrosterone
	Adrenal Cortex	Cortisol, Corticosterone, 11-deoxycoricosterone, 11-deoxycorisol, aldosterone, 18-hydroxycorticosterone, 18-hydroxy-11-deoxycorticosterone
	Corpus Luteum	Progesterone
Peptide hormone	Hypothalamus	Corticotropin-releasing factor, Follicle-stimulating-hormone-releasing factor, Gonadotropin-releasing factor, Luteinizing hormone-releasing factor, Melanotropin-releasing factor, Melanotropin release-inhibiting factor, Prolactin-releasing factor, Prolactin release-inhibiting factor, Somatotropin-releasing factor; growth hormone-releasing factor, Somatostatin (Somatotropin release-inhibiting factor), Thyrotropin-releasing factor
	Pituitary	Chorionic gonadotropin, Chorionic somatomammotropin, Adrenocorticotropic hormone, Follicle-stimulating hormone, Gonadotropin hormone, Oxytocin, Lipotropic hormone, Luteinizing hormone; (Interstitial cell-stimulating hormone), Melanocyte-stimulating hormone, Prolactin, Somatotropin (Somatropic hormone; growth hormone), Thyrotropic hormone, (Human) Menopausal gonadotropin, Vasopressin (Adiuretin; antidiuretic hormone)
	Pancreas	Insulin and Glucagon
	Parathyroid	Parathormone/Collip's Hormone/Parathyroid Hormone
	Gastrointestinal Tract	Gastrin, Secretin, Cholecystokinin, Pancreozymin, enterogastrone, Enterkrinin, Hepatocrinin, Duicrinin, Villikinin, Parotin
	Corpus Luteum	Relaxin
	Thyroid	Calcitonin
Hormones as amino acid derivatives	Thyroid	Thyroxine and 3, 5, 3'-triodothyronine
	Adrenal medulla	Epinephrine (adrenaline) and norepinephrine (noradrenaline)

Enzymatic cleavage (endopeptidase and post proline carboxyamide endopeptidase) at Tyr^5-Gly^6 and Pro^9-Gly^{10} degrades GnRH. The structure-function relationship study of GnRH has shown that changes made at Gly^6 with specific D-amino acids and at C-terminal of the peptide enhance the function of GnRH analogs reducing its predisposition towards the enzymatic cleavage [126].

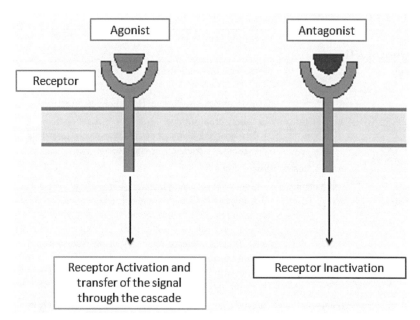

FIGURE 2.43   Agonist and antagonist.

The GnRH binds to GnRH receptors in the pituitary (belongs to G-protein coupled receptors family) and activates them. The activated receptors result in breakdown of phosphoinositide producing inositol 1,4,5-trisphosphate $(Ins(1,4,5)P3)$ and diacylglycerol. The release of Ca^{2+} ions from intracellular stores and activation of protein kinase C is initiated by secondary messengers. If the GnRH receptors are desensitized on persistent of the receptors. Hence, this results in suppression of secretion of gonadotropins.

GnRH agonists (GnRHa) are those molecules which mimic the function of GnRH and stimulate release of LH and FSH. A continuous treatment with GnRHa results in down-regulation of LH and FSH production [130]. The continued use of GnRH superagonists results in drop of the levels of steroid hormones giving rise to no fertility conditions in males and females. The continuous use of GnRHa is therefore beneficial in cases like precocious puberty, endometriosis, and advanced metastatic breast and prostrates cancers [126]. The GnRH antagonists (GnRHag) function by acting as competitive inhibitors for the GnRH receptors. The GnRHag can be used to prevent surge of LH and thus ovarian hyperstimulation syndrome [130]. The list of GnRH agonist and antagonist is given in Table 2.13.

TABLE 2.13   List of GnRH Agonist and Antagonists

Agonist	Antagonist
Leuprolide acetate	Ganirelix actate
Goserelin acetate	Cetrorelix
Nafarelin acetate	Ganlrelix
Histrelin acetate	
Triptorelin pamoate	

2.26.2   HORMONES OF PITUITARY

Pituitary gland is present below the brain the fossa of sphenoid bone called sella turcica. The pituitary gland has two lobes: an anterior lobe, adenohypophysis (comprise 2/3rd of the gland) and a posterior lobe, neurohypophysis (comprises 1/3rd of the gland) [131]. A small intermediate lobe is present between the anterior and the posterior lobes and is referred to as pars intermedia [1]. The secretions of pituitary gland regulate the functioning of adrenal cortex, sex organs, and thyroid. The hormone secretion of adenohypophysis is regulated by the hyptothalamic hormones. The neurohypophysis secretes and stores oxytocin and vasopressin. The release of these hormones is also under the control of hypothalamus. The hormones of pituitary and the peptide drugs are described below.

2.26.2.1   SOMATOTROPHIN (ST) (GROWTH HORMONE)

Somatotrophin (ST) is a polypeptide composed of 191 amino acids. ST comprises of two disulfide bonds and four helices which form the part of three-dimensional structure of protein. The ST is secreted by the pituitary under the influence of growth hormone-releasing hormone (GHRH). The ST secretion is inhibited by the hormone somatostatin.

The functions of ST are as follows:

- Stimulates growth of bones;
- Increase total mass and mineral content of bones;
- Increase lean body mass and decreases adipocity;
- Growth in size and function of the organs;
- Maintains the function of pancreatic islets.

Many activities of ST are mediated by the synthesis of insulin-like growth factors-1 and -2 (IGF-1 and IGF-2) [131].

The ST binds to ST receptors (STR) resulting in the dimerization of the receptors followed by the activation of JAK2-tyrosine kinase associated with STR. The JAK2-tyrosine kinase phosphorylates the tyorine residues of JAK2 as well as STR [132]. The STAT kinases are phosphorylated by phosphorylated JAK2. The phosphorylated STAT kinase dimerises and moves into the nucleus. Inside the nucleus, they activate transcription factors by phosphorylation and hence transmit the signal down the cascade resulting in synthesis of various proteins [131].

The ST is responsible for skeletal growth in children. The absence of ST during childhood results in dwarfism. Earlier, the dwarfism was treated by administering human growth hormone (hGH) obtained from cadavers. The use was discontinued by the US FDA due to the death of children after the administration of hGH. The deaths occurred due to Creutzfelt Jakob disease caused by prions. Today, the dwarfism is treated by administering the recombinant hGH [126].

2.26.2.2   ST ANALOGS AND ANTAGONISTS

2.26.2.2.1   Recombinant hGH (rhGH)

The rhGH is used for the treatment of conditions like growth hormone deficiency (GHD), Turner syndrome (TS), Prader-Willi syndrome (PWS), chronic renal insufficiency (CRI), short stature homeobox-containing gene deficiency (SHOX-D) and being born small for gestational age (SGA) all of which are characterized by short stature [133]. Adults suffering from ST deficiency are also benefitted by treatment with rhGH replacement therapy. rhGH is also useful in weight loss and cachexia occurring as a result of AIDS and in patients with high severity of burns [126].

2.26.2.2.2   ST Antagonists

The excessive synthesis of ST results in a condition called acromegaly. The major reasons behind the excessive synthesis of ST are somatotroph adenomas of the pituitary gland. The acromegaly can be treated by removing the adenomas surgically followed by radiation therapy and administering drugs. Somatostatin analogs like octreotide are used for the inhibition of growth hormone synthesis. These analogs also inhibit the secretion of insulin, glucagon, and several GI hormones and trigger cholelithiasis. Pegvisomant is a recombinant hGH analog which has an ST receptor antagonist [134].

The Pegvisomant is 191 amino acid molecules with changes in amino acid residues at 18, 21, 120, 167, 168, 171, 172, 174, and 179 are covalently linked to polyethylene glycol. The Pegvisomant binds to STR and hence inhibits the signaling cascades resulting in no synthesis of IGFs and hindering the problems associated with acromegaly [126].

2.26.2.3   GONADOTROPHINS

The two gonadotrophins produced by adenohypophysis are LH and FSH. Both the LH and FSH are glycoproteins made of α and β subunit [135]. The α subunit of both the proteins are the same and composed of 92 amino acids. The β subunit of LH has 121 amino acids and that of FSH has 111 amino acids. The biological activity of these hormones can be attributed to the β subunits [126].

The LH and SH are secreted by pituitary on stimulus by GnRH. The LH and FSH are glycosylated in the secretory cells of pituitary. The LH and FSH reach the target cells and bind to receptors on the cell surface. The receptors of LH and FSH are G-linked proteins. Activation of the receptor by binding of the hormones stimulates adenylate kinase to function resulting in subsequent increase in concentration of cAMP. The signaling cascade also involves activation of phospholipase C and hydrolysis of phosphoinositide [135]. The functions of LH and FSH are given in Table 2.14.

TABLE 2.14   Functions of LH and FSH

	FSH	LH
Females	Maturation of ovarian follicle	Ovulation
	Secretion of estradiol	Corpus luteum formation
		Progesterone secretion
Males	Maturation of sperms in testes	Secretion of testosterone

The functions of LH and FSH are involved in the fertility of males and females and hence they are available commercially for various therapeutic usages. The various peptide drugs are described below:

- **Menotropins:** They are equimolar mixture of FSH and LH (1:1). It is isolated from the urine of woman in post-menopause. The functions of menotropins are given in Table 2.15.

- **Follitropins:** They are hormonal products which purely consist of FSH [126]. Follitropin α (Fα) is a recombinant FSH. Fα is used for development of ovarian follicle sin women undergoing IVF treatment [136]. Follitropin β is very similar to Fα and are approved for medical use [126]. The function of follitropins is given in Table 2.15.
- **Urofollitropins:** They are also obtained from the urine of women in post-menopause followed by purification to yield FSH with insignificant qunatities of LH. The functions of urofollitropins are described in Table 2.15.
- **Lutropin α:** It is an analog of LH and is synthesized by recombinant DNA technology [126]. The functions of lutropin are listed in Table 2.15.

TABLE 2.15 Peptide Drugs Related to Gonadotrophins

Name of the Drug	Functions
Menotropins	Stimulate spermatogenesis in males with primary or secondary hypogaonadisms
	Increase testosterone secretion
	Induction of ovulation in females
	In combination with human chorionic gonadotropin (hCG)
	Promotes development of multiple follicles in women undergoing IVF treatment
Follitropins	Stimulates growth of ovarian follicles
Urofollitropins	Stimulates development of ovarian follicle
	Used in treatment of infertility arising because of polycystic ovarian syndrome
Lutropin α	Used for treatment of women in assisted reproductive technology
	In combination with Follitropin α stimulates growth of follicels in women

2.26.2.4   ADRENOCORTICOTROPHIC HORMONE (ACTH)

Adrenocorticotrophic hormone (ACTH) is a hormone regulating the synthesis and release of hormones of adrenal cortex. The ACTH is produced from the precursor pro-opiomelanocortin (POMC) by action of peptidases. POMC also produces opioid peptides and melanocyte-stimulating hormones under the action of enzymes. The receptor for ACTH is a G-linked protein and cAMP is the secondary messenger. Excess synthesis of ACTH results in Cushing's syndrome [138].

ACTH is a peptide composed of 39 amino acids. The amino acids 1–24 from the N-terminal are responsible for the biological activity of the hormone. The remaining 15 amino acids are species-specific and protect the hormone from proteolytic cleavage. ACTH is commercially obtained from natural sources. It is administered in the form subcutaneous (SC) or intramuscular (IM) injection. It is also used in treating acute exacerbations in multiple sclerosis due to its anti-inflammatory and immunosuppressive potential [126].

Cosyntropin is a synthetic analog of ACTH. It is also referred to as tetra-cosactide. The 1–24 amino acid residues of cosyntropin are the same as that of ACTH [140]. Cosyntropin is also used in the ACTH stimulation test to check the function of adrenal and pituitary [138].

2.26.2.5 OXYTOCIN

Oxytocin is a cyclic peptide secreted by neurohypophysis. Oxytocin performs two major functions:

- Contraction of muscles of uterine during parturition; and
- Release of milk from the mammary gland by contraction of myoepi-thelial cells surrounding the alveoli of the mammary gland [140].

Synthetic oxytocin is also referred to as Pitocin or Syntocinon and is used to induce labor. It is also immediately after delivery to deal with post-partum hemorrhage. Oxytocin is also used in the form of nasal spray to aid young mothers in lactation [141].

2.26.2.6 VASOPRESSIN (ARGININE-VASOPRESSIN, AVP)

Vasopressin is a cyclic peptide very similar to oxytocin [142]. It is also referred to as [Phe^3, Arg^8] Oxytocin. The functions of vasopressin are as follows:

- Reabsorption of water from renal tubules;
- Increasing blood pressure by contraction of arterioles and capillaries [126].

Deficiency of vasopressin results in diabetes insipidus. This condition is characterized by polyuria and polydipsia [140].

Desmopressin [DDAVP; 1-deamino-8-D-arginine vasopressin], has N-terminal cysteine devoid of α amino group and L-arginine at 8th position form n-terminal replaced by its D-form the molecule has an increased half-life and hence can be used through oral, parenteral, and nasal routes. It is found to be effective in diabetes insipidus and nocturnal enuresis [126].

DDVAP is also a well-accepted hemostatic agent and can cause the release of coagulation factor VII, von Willebrand factor, and tissue-type plasminogen activator [142]. Hence, it is used to treat patients with Hemophilia A and type I von Willebrand disease [126].

2.26.3   HORMONE OF PARATHYROID GLAND

The parathyroid gland is involved in the secretion of parathormone (PTH). There are four parathyroid glands located such that a pair lies above and below the posterior surface of each lobe of the thyroid gland [126]. Parathyroid hormone is a peptide 84 amino acids secreted by the chief cells of the gland. The biological activity of the hormone is conferred in the first 34 amino acids of the N-terminal. Removal of N-terminal and C-terminal amino acids produces the non-functional truncated hormone. Removal of the 1st and 2nd amino acids from the N-terminal results in retention of receptor binding capacity but reduced biological activity. PTH has a very short half-life of 4 minutes. PTH performs various functions as described below:

1. **In Bones:** Osteolysis, osteoclasts differentiation, remodeling of bone, and bone resorption.
2. **In Kidneys:** Reabsorption of calcium ions (Ca^{2+}), phosphate resorption is inhibited and 25–OH hydroxylation of vitamin D [143].

The deficiency of PTH results in hypoparathyroidism characterized by hypocalcemia and hyperphosphatemia.

Recombinant PTH-Teriparatide is used for the treatment of hypoparathyroidism [144]. The first 34 amino acid of teriparatide is the same as that of PTH. Teriparatide can be administered subcutaneously and it enhances osteoblastic activity, promoting bone formation. It is used for treating hypogonadal osteoporosis and postmenopausal osteoporosis [126].

2.26.4   HORMONE OF PANCREAS

The pancreas is located behind and below the stomach transversely. The pancreas is both an exocrine and an endocrine gland. The pancreas is made of acini (exocrine in function) and islets of Langerhans (endocrine in function). Islets of Langerhans have three types of cells:

*   α cells which produce glucagon;
*   β cells which produce insulin; and

- δ cells which produce somatostatin [1].

The hormone insulin and related peptide drugs are described in subsections.

2.26.4.1   INSULIN

Insulin a hormone composed of 51 amino acids. It consists of two chains: A and B, linked by disulfide bonds. The chain A has 21 amino acids and chain B has 30 amino acids. The gene for insulin produces protein pre-proinsulin. A 23 fragment of the pre-proinsulin is cleaved off during translation as the growing peptide enters the ER. The chain A and B are connected by peptide C which is essential for the formation of disulfide bonds. This form of insulin with chain A and B and peptide C is referred to as proinsulin. The peptide C is cleaved off in the Golgi apparatus forming a functional insulin molecule (chain A and B linked by two disulfide bonds). Insulin is secreted by the pancreas in pulsatile manner every 10 minutes. It has a short half-life of 3 minutes in systemic circulation. The various functions of insulin are as follows:

- Uptake of amino acids by muscles and promotion of protein synthesis;
- Prevention of protein breakdown [145];
- Uptake of glucose by muscles and liver and its storage in the form glycogen [126].

Diabetes mellitus (DM) results because of increased concentration of glucose in blood [145]. DM is characterized by hyperglycemia, glycos-uria, ketonemia, ketonuria, polyuria, nocturia, polydipsia, polyphagia, and asthenia. There are two types of DM:

1. Type I DM (insulin-dependent DM) resulting due to destruction of β cells of islets of Langerhans. Thus very little or no insulin is produced by the body.
2. Type II DM (Noninsulin dependent DM) resulting due to failure of recognition of insulin by the insulin receptors arising because of insulin resistance [1].

The insulin analogs are utilized for treating DM. There are two types of insulin analogs:

i. **Rapid-Acting Analogs:** They are fast-acting short-lived analogs; and

ii. **Long-Acting Basal Analogs:** They are slow at the start of action, released slowly and continuously in the systemic blood circulation. Thus, they mimic normal insulin basal levels and aid in control of basal glycemia and nocturnal hypoglycemia [146].

The insulin analogs with their characteristic features are given in Table 2.16.

TABLE 2.16   Insulin Analogs and its Characteristic Features

Type of Analog	Names of Analog	Structural Features	Functions
Long-acting analogs	Glargine	Asparagine at position 21 is substituted by glycine in chain A and elongation of C-terminal of B chain by two arginine residues	Onset of action after 2 hours of injection; decrease in hypoglycemic events and low risk of nocturnal hypoglycemia
	Detemir	Acylation of myristic acid to the lysine residue at position 29 and deletion of threonine 30 in chain B	Extended pharmacodynamics profile with functions lasting for about 17 hours
Rapid-acting analogs	Lispro	Genetically engineered; approved for clinical use in 1996; reversal of proline at position 28 and lysine at position 29 in B chain	Improves The postprandial leptin and grehlin in type I DM May be used in gestational diabetes
	Aspart	Substitution of proline at position 28 by aspartic acid in B chain	Better control of glycemia if administered before meal
	Glulisine	Asparagine at position 3 is substituted by lysine and lysine at position 29 by glutamic acid in chain B; launched in 2004	Most fast acting analog

2.27   PEPTIDE AND PROTEIN DRUG DELIVERY METHODS

Peptides and proteins are enzymes, hormones, structural molecules, etc. The proteins perform numerous functions in the body of an organism and therefore possess tremendous potential for use in therapeutics. The peptide and protein-based therapeutics have increased in the recent years because of the following reasons:

• Better analytical techniques leading to discovery of peptides and proteins useful in medication;

- Increased production of the peptides and proteins due to genetic engineering and molecular biology techniques;
- Improved perception of peptides and proteins in the pathophysiology of diseases [147].

The delivery of the drug in the body is not very simple. The peptide or protein drugs have to cross hydrophilic and lipophilic barriers to reach the target cell in the body of the organisms. There various limitations posed to drug deliveries which are described as follows:

- Barriers posed by enzyme limit the absorption by GI tract.
- Transport of drugs across the intestinal epithelium.
- Transport of drug across the endothelium of the blood capillaries.
- Transport of drugs to the brain across the blood-brain barrier.

The peptide and protein drugs are delivered in various ways according to the following classification [148] (as shown in Figure 2.44):

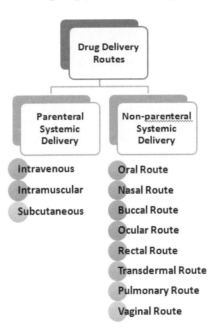

FIGURE 2.44   Classification of drug delivery methods.

1. **Parenteral Systemic Delivery:** The parenteral route of drug delivery is the direct delivery of the drug in the systemic circulation. This mode of drug delivery is commonly used for drugs that are weakly

absorbed by the GI tract. This mode of drug delivery is also useful for patients who are unconscious or require immediate rapid action of drug. The use of this route also helps control the quantity of the drug administered to the individual. The parenteral route of drug delivery is irreversible in nature and can cause pain, fear, and damage to local tissue or infection. The parenteral drug delivery is of three types:

i. **Intravenous (IV):** IV injection is the most commonly used route of drug delivery. It is a rapid method of delivering the controlled amount of drug in the systemic circulation. The advantages and disadvantages of the IV method of drug delivery are given in Table 2.17 [149]. IV method is the second preferred method of parenteral drug delivery for peptide and protein drugs [148].

TABLE 2.17   Different Types of Parenteral Routes, Advantages, and Disadvantages

Route of Administration	Advantages	Disadvantages
Intravenous (IV)	Useful for chemicals which may irritate GI tract	Cannot be recalled by binding strategies like using activated charcoal
	Drug is rapidly diluted by blood	May result in infection due to introduction of bacteria or other pathogens at the site of infection
	Immediate delivery of the drug in the systemic circulation	May induce hemolysis
Intramuscular	Continuous slow release of drug	Release of drug after diffusion of vehicle
		No direct delivery in systemic circulation
Subcutaneous	Reduced possibility of hemolysis or thrombosis	Cannot be used for drugs that may cause irritation to tissue
	Continuous and slow release of drug	

ii. **Intramuscular (IM):** IM route of delivery is used for drugs available as depot preparations (suspension of drug in non-aqueous vehicle like polyethylene glycol) and absorbed slowly. The vehicle diffuses out of the muscle followed by the precipitation of drug. The drug dissolves slowly and is released in continuous manner [149]. For example, IM administration of γ globulin for protection from hepatitis [150].

iii. **Subcutaneous (SC):** SC route of drug administration is slow than IV mode. This mode provides continuous and slow release of drug with reduced possibility of hemolysis or thrombosis [149]. Insulin analogs, LH releasing hormone analog (goserelin), LH releasing hormone agonist (nafarelin), etc., are administered by SC route [148].

2. **Non-Parenteral Systemic Delivery:**
   i. **Oral Route:** The oral route of drug administration is one of the most common drug delivery routes because of following reasons:

   - Self-administration of drugs;
   - Decreased chances of systemic infections;
   - Drug dose can be reversed by use of activated charcoal.

   The oral route also possesses certain disadvantages like:

   - Complicated absorption;
   - Inactivation of drug due to low pH in stomach [149].

   The delivery of peptide and protein drug faces many problems when administered by oral route: extreme pH conditions, protease degradation, decreased bioavailability, and tendency to undergo aggregation, short half-life in plasma, traversing the epithelium, adsorption, and denaturation. Due to such constraints, only two protein-based drugs – interferon-α and human growth hormone – are orally administered [150]. Various approaches used for oral drug delivery are explained in Table 2.18.

   ii. **Nasal Route:** Nasal route of drug delivery, delivers the drug directly into the systemic circulation via lungs [150]. The drug delivery via nasal route possesses certain advantages and disadvantages listed in Table 2.19 [148].

   The various problems of peptides and proteins drugs administration by nasal route are low bioavailability, large molecular weight (>1000 Da), passive diffusion, irritation of respiratory mucosa and hydrolytic/proteolytic degradation. The peptide and protein drug delivery can be improved by modifying pH and viscosity of the drug, preventing protein aggregation, increasing blood flow to nasal region and using penetration enhancers. Peptides and proteins that are delivered by nasal route are oxytocin and nafarelinactate [150].

TABLE 2.18 Strategies for Oral Drug Delivery of Peptides and Proteins

Strategy Available	Types of Strategy	Features
Chemical Modification	Prodrug approach	Proteins are modified to increase their half-life and stability in plasma.
		Modification is done by olefenic substitution, carboxyl reduction dehydroamino acid substitution, d-amino acid substitution, thiomethylene modification, attaching PEG and retro inverso modification.
	Nobex technology	PEG and alkyl groups or fatty acid radicals are linked to the amino group of protein, resulting in amphiphilic molecules (AM).
		AM can travel across the mucosa and resist enzymatic cleavage.
		AM have decreased association with enhanced penetration and compatibility.
Protease Inhibitors	—	Co-administered with the protein drug.
		Enhance the stability of drug.
		For example, chymotrypsin is inhibited by chymostatin; endoprotease is inhibited by α_2 macroglobulin, etc.
Penetration Enhancers	—	Protein molecules are large and cannot easily traverse across the plasma membranes to reach systemic circulation.
		Surfactants, bile salts, Ca^{2+} chelating agents, fatty acids, alkanoylcholines, chitosans, phospholipids, etc., function as penetration enhancers.
		Provide protection from proteases of intestine.
		Proteins are stabilized against denaturation due to decreased self-association and absorption on hydrophobic interface of matrix used for delivery.
Formulation Approach	Emulsions	Protects from acid and luminal proteases of GI tract.
		Increased permeation through mucosa of intestine.
	Microspheres	Prevents proteolytic degradation in stomach.

TABLE 2.18   *(Continued)*

Strategy Available	Types of Strategy	Features
	Liposomes	Stability of protein is improved.
		Better permeability through the membrane.
	Nanoparticles	Recent and advanced approach.
		Decreased susceptibility to degradation by enzymes.
		Improved absorption by epithelium of intestine.
Mucoadhesive polymeric system	—	Used for drug delivery at specific site.
		Better permeability through the plasma membrane.
		Improved bioavailability.
		For example, polyacrylic acid or cellulose derivatives include.
		Carbopol, polycarbophilpolyacrylic acid (PAAc), poly(2-hydroxyethyl methacrylate), poly(methacrylate), carboxymethyl cellulose, hydroxyl ethyl cellulose, sodium carboxymethyl cellulose, methylcellulose, methyl hydroxyethyl cellulose, etc.

TABLE 2.19   Advantages and Disadvantages of delivering active pharmaceutical agents through nasal route

Advantages	Disadvantages
Easy way of administering drug	Long term usage can harm the respiratory mucosa
Better absorption due to rich vascularization	The environment of the nasal passage is modified in conditions like common cold, etc.
Avoidance of first pass metabolism	
Quick action of drug	

iii. **Buccal Route:** It is one of the recently developed routes of drug delivery. The buccal route of drug delivery has advantages given as follows:

- Easily accessible;
- No degradation of proteins;
- Does not directly enter the hepatic metabolism;
- Drugs can be delivered with the aid of patch.

Different strategies have been employed for successful drug delivery through the oral route. They are described as follows:

- Use of adhesive gels;
- Use of adhesive patches;
- Use of adhesive promoters;
- Use of adhesive tablets.

Peptides that have been used through buccal route include oxytocin, vasopressin, LHRH analogs, calcitonin, insulin, etc. [148].

iv. **Ocular Route:** The use of ocular route of drug delivery can help deliver drug into the systemic circulation. The drug can be introduced in the precorneal cavity reaches systemic circulation through the blood vessels underlying the mucosa of conjunctiva or overflow in the nasolacrimal drainage system resulting in absorption by nasal mucosa [150]. The absorption of the peptide or protein drugs can be enhanced by the use of use of nanoparticles, liposomes, gels, ocular inserts, bioadhesives, or surfactants [148]. The absorption of drugs can be reduced through the ocular route due to dilution by tears, lachrymal drainage, etc. [150].

v. **Rectal Route:** The drugs delivery through the rectal route is distinct as the drug introduced through this route does not enter the portal system of blood circulation. As the drug does not enter the GI tract, its destruction is prevented from the enzymes and adverse pH conditions. It is also beneficial in case of drugs which induce vomiting and also if the patient has lost his/her consciousness. The disadvantage of the rectal route of administration is irritation of rectal mucosa by drugs [149].

The use of rectal route for peptide and protein drug delivery is a very recent approach. The insulin co-administered with sodium 5-methoxy salicylate through microenema has been reported in humans [150]. Besides, various peptides and proteins which have been studied for rectal route of drug delivery include Vasopressin and its analogs, pentagastrin, and gastrin, calcitonin analogs and human albumin [148].

vi. **Transdermal Route:** The absorption of drugs through the skin comprises the transdermal route. The transdermal route of drug delivery is used for nitroglycerin, scopolamine, and nicotine for easing the process of quitting smoking [149]. The peptide and protein drugs through the transdermal route is a recent approach of drug delivery. The surface of the skin possesses aminopeptidase which have a lower proteolytic potential than the enzymes of the GI tract. The bioavailability of the drug is comparatively greater than the oral route of drug administration. The protein-drug permeation is increased by the use of mechanical abrasion and chemical enhancers [150]. The various approaches used for drug delivery through the transdermal route are described in Table 2.20.

TABLE 2.20   Strategies for Transdermal Drug Delivery of Peptides and Proteins

Strategy Used	Features
Iontophoresis	Non-invasive technique.
	Used for local and systemic delivery of drugs.
	Charged protein molecules are transported across the membrane.
	Proteins are charged by changing the pH and ionic strength of the solution.
	Heat generated during this process can denature proteins.
	Has been studied for insulin, salmon calcitonin, etc.
Phonophoresis	Ultrasound is applied to the skin via coupling contact agent.
	Thermal effect of ultrasonic waves increases drug absorption.
	Peptides or proteins can get denatured due to increased temperature or mechanical disruption.
Penetration enhancers	Useful in topical delivery of drugs.
	Oleic acid, dimethyl sulfoxide, surfactants, and azone have been used as penetration enhancers.
	Skin irritation is a major disadvantage.
Prodrugs	Peptides and proteins are modified to form prodrugs.
	LHRH, TRH, and neurotensin have been delivered using this approach.

vii. **Pulmonary Route:** Pulmonary route of drug delivery involves delivering the drugs to the lungs. The drugs delivered to lungs can be easily absorbed due to transcytosis. The lungs provide large surface are for absorption. The devices that are available to deliver drugs to lung are:

• Metered dose inhaler;
• Nebulizer;

- Powder inhaler/Insufflator.

The advantages and disadvantages of pulmonary route of drug delivery are given in Table 2.21.

TABLE 2.21   Advantages and Disadvantages of Nasal Route of Peptide and Protein Drug Delivery

Advantages	Disadvantages
Straight entry of drug into blood circulation	Drug delivered generally to upper part of the lungs characterized by low absorption
Dose requirements reduced by about 50 times	Small quantities of drug can be delivered
Quick absorption	
Suitable for patients with lung diseases	
Does not elicit immune reaction	
Least discomfort and pain to patient	

There are different approaches for drug delivery through the pulmonary route. They are described in subsections.

2.27.1   PARACELLULAR DELIVERY

The paracellular delivery of drugs is a novel approach of drug delivery. The drugs are delivered through the water-filled pathway. The drugs travel in the space between two cells, avoiding contact with lysosomal enzymes. The major limitation of this method is small pore size of the pathway which may be regulated by cell signaling pathways.

2.27.2   SMART POLYMER-BASED DELIVERY

Injectable smart polymer-based drug delivery is an interesting method. Biodegradable polymers may be used for delivering peptide and protein drugs. This system has various advantages like easy application, site-specific drug delivery, sustained drug delivery, sustained drug delivery, reduced drug dose, and better comfort for patient.

These polymer-based drugs can be easily prepared. They form implants at the injection site. Smart polymers are classified based on the response to various stimuli. The classification of smart polymers with examples is given in Table 2.22.

TABLE 2.22 Classification of Smart Polymers

Type of Smart Polymer	Examples
Temperature sensitive polymer	Poly(ethylene oxide)-poly(propylene oxide)-poly(ethylene oxide) triblock copolymers (PEO-PPO-PEO) and Poly nisopropylacrylamide (PNIPAAM)
Phase-sensitive polymer	Poly(D,L-lactide) and Poly(D,L-lactide-co glycolide)
pH-sensitive polymer	Poly(methacrylic acid g-ethylene glycol) P(MAA-gEG)
Photo-sensitive polymer	PEG, Poly(vinyl alcohol), and PEO-PPO.

2.27.3 HYBRID PROTEIN DELIVERY

Hybrid proteins comprising properties of two or more protein can be used for site-specific drug delivery using techniques like gene ligation (hybrid protein has been developed by fusing Interferon γ and TNF β) or linking protein fragments by chemical bonds [148].

- **Vaginal Route:** Vaginal route of drug delivery is useful for prolonged systemic medication. The various advantages of this method are as follows:

    – Ease of self-administration;
    – Extended retention of drug;
    – Reduced proteolytic degradation.

The major disadvantage of this method is variable permeability of vaginal surface due to variation in serum estrogen levels. LHRH and analogs can be administered using this mode of drug delivery [150].

2.28 METHODS OF PEPTIDE SYNTHESIS

Peptide synthesis is important for manufacture of peptides in industries. Many methods of peptide synthesis have been developed. Some of the methods of peptide synthesis are [152]:

1. Solid-phase peptide synthesis (SPPS);

2. Solution phase synthesis (SPS);
3. Native chemical ligation (NCL);
4. Staudinger ligation (SL);
5. Click chemistry (CC).

The various methods of peptide synthesis are described in subsections.

2.28.1 *SOLID PHASE PEPTIDE SYNTHESIS (SPPS)*

The SPPS method was introduced by Merrifield in 1963. This method makes use of resin as support [152]. The amino group of amino acid to be linked to the resin is protected by temporary protecting group which is generally a urethane derivative [151]. The protected amino acid links to amino group of the resin by its C-terminal. Addition of the protecting group prevents polymerization. The commonly used protecting groups are Fmoc (9-fluorenyl-methyl carbamate) and t-Boc (Di-tert-butyl dicarbonate). After the addition of amino acid to the resin, protection group is removed using trifluoroacetic acid (for t-Boc) or 20% piperidine in dimethylformamide (for Fmoc) and resin is washed before the next addition. The process is repeated using the amino-terminal of amino acid attached to the resin in order to link the next amino acid in the sequence. The new amino acid which is added also has its amino group blocked by protecting group to prevent polymerization. These steps are repeated again and again to obtain desired sequence of peptide. Resins used as support for SPPS are polystyrene, Merrifield, hydroxymethyl, phenylacetamidomethyl, Wang, and 4-methylbenzhydrylamine resins. The SPPS is shown in Figure 2.45.

Microwave-assisted SPPS is an improved SPPS technique used for synthesis of long peptides. The peptides are produced in high yield with low racemization. This method also enables better control of temperature and pressure during the synthesis process [152].

➢ **Advantages:**

• Excess of reagents and side products can be separated from the growing peptide.
• All the steps of the process can be performed in a single reaction vessel [151].

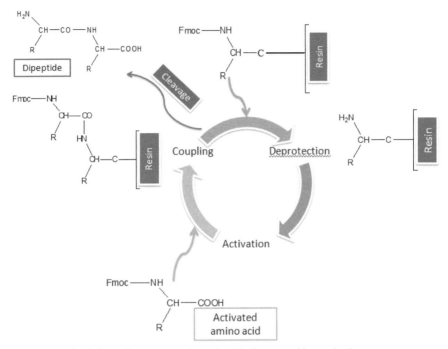

FIGURE 2.45   Schematic representation of solid-phase peptide synthesis.

➢ **Disadvantages:**

- Resins and equipment used are very expensive [152].

2.28.2   *SOLUTION PHASE SYNTHESIS (SPS)*

The solution-phase synthesis (SPS) method is centered on the concept of coupling of amino acids in solution. Long peptides are synthesized by using fragment condensation method. Short fragments of the required peptide are synthesized followed by joining of these short peptides to give long peptide.

➢ **Advantages:**

- Intermediated products can be deprotected and purified, hence, producing required peptide in pure form.
- Scaling up of the process is easy.

➢ **Disadvantages:**

• Extended reaction time.

The SPS method is used to produce oxytocin, human insulin, etc. [152].

2.28.3 CHEMICAL LIGATION (CL)

In this technique, small peptides are selectively coupled to from large peptides. The two peptides are coupled together by thiol capture method. Thiol capture involves formation of disulfide bond between cysteine residue of N-terminal and thiol group of C-terminal of the peptides. The thiol group is then replaced by an acyl group to form a peptide bond [152].

2.28.4 NATIVE CHEMICAL LIGATION (NCL)

The technique of NCL was developed due to optimization of CL. This technique was developed by Dawson in 1994. Interluekin 8 has been produced by this technique. A large peptide is produced by linking N-terminal cysteine of one unprotected peptide with α-thioester group of another unprotected peptide forming an intermediate containing thioester moiety. Intramolecular acyl migration converts this thioester linkage to a peptide bond. This reaction is carried out in an aqueous buffered solution at neutral pH. The process of NCL is represented in Figure 2.46.

➢ **Advantages:**

• Starting materials are very stable;
• Large chemoselective nature;
• Unprotected ends are used for ligation;
• Can be used for producing peptide thioesters.

2.28.5 STAUDINGER LIGATION (SL)

The Staudinger ligation(SL)technique was developed by Raines and Bertozzi. The SL technique is an alternative technique for NCL. It is based on the reaction of a phosphane with an azide to form an iminophosphorane. The iminophospjorane undergoes an acyl shift to form amidophosphonium salt. The hydrolysis of amidophosphonium salt yields an amide and phosphine oxide. This method can be used for site-selective modification of protein as well as for peptide and protein immobilization. The schematic representation of the process is given in Figure 2.47.

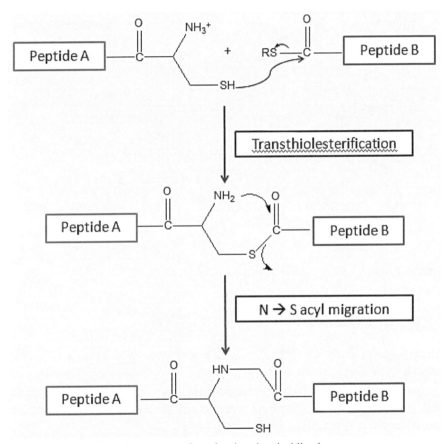

FIGURE 2.46   Schematic representation of native chemical ligation.

2.28.6 CLICK CHEMISTRY (CC)

This technique is developed by Sharples in 2001. It is based on copper(I) catalyzed cycloaddition of alkynes and azides to form 1,2,3-triazoles. It is commonly used process for drug development and discovery [152].

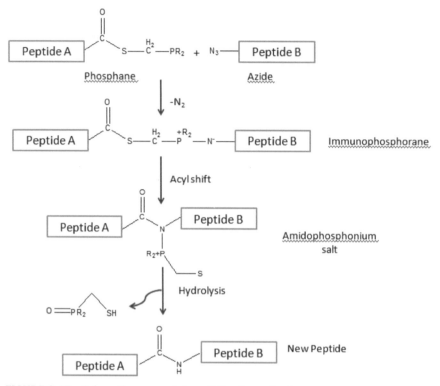

FIGURE 2.47 Schematic representation of Staudinger ligation.

2.29 PEPTIDES AND PROTEIN MEDIATORS

Molecules involved in cell-cell signaling are referred to as mediators. Mediators perform diverse roles in the process of cell signaling. Peptides and protein are important mediators and perform many important physiological and pathophysiological functions. Peptide and protein mediators with their classification and functions are listed in Table 2.23 [153].

2.30 CONCLUSION

This comprehensive and interesting book chapter on proteins and peptides exclusively demonstrated the basic traditional aspects, complete knowledge of the essential and non-essential amino acids (Dissociation Constant, Density, Specific Rotation, Melting Point, Boiling Point, and Isoelectric Point), structural composition, physical attributes (shape, size, molecular weight, color, taste, solubility, denaturation, curdling, etc.), chemical features (acid hydrolysis, alkaline hydrolysis, enzymatic hydrolysis, benzylation, formylation, acylation, oxidation, decarboxylation, esterification, reduction, etc.), optical properties, biological importance (hormones, etc.), nomenclature, classification (based on shape of the protein, based on composition and solubility of the protein, and based on function of the protein), peptide bond properties, specific functions, hierarchy of protein structure (primary structure of protein, secondary structure of protein, tertiary structure of protein, and quaternary structure of protein), structure-function relationships, techniques for protein structure determination (X-ray crystallography, NMR spectroscopy, etc.), quantitative estimation (bichinchonic acid, Bradford method, Lowry method, etc.), delivery systems (parenteral systemic delivery, non-parenteral systemic delivery), methods of peptide synthesis (solid-phase peptide synthesis, click chemistry, native chemical ligation, solution phase synthesis, Staudinger ligation, etc.), etc. which will positively motivate the young minds and researchers of diverse fields(life sciences, pharmaceuticals, chemical sciences, etc.) across the globe in enhancing their knowledge and translating them into useful discoveries and therapeutically privileged products.

TABLE 2.23   Peptide and Protein Mediators with Their Function

Type of Peptide Mediators	Example of Peptide/ Protein Mediator	Function
Neurotransmitters and neuroendocrine mediators	Neuropeptide Y	Vasoconstriction
Hormones form non-neural sources	Angiotensin II	Increase blood pressure by vasoconstriction of blood vessels
	Bradykinin	Plays an important role in inflammation
Growth factors	Insulin-like growth factors (IGFs)	Cell proliferation
Mediators of immune system	Cytokines	Involved in inflammation

KEYWORDS

- **amino acids**
- **chemical reaction**

- **peptide synthesis**
- **peptides**
- **protein delivery**
- **proteins**
- **therapeutics**

REFERENCES

1. Jain, J. L., Jain, S., & Jain, N., (2007). *Fundamentals of Biochemistry*. S. Chand and Company Ltd.
2. Linda, J., Guro, G., & Elias, S. J. A., (2005). Selenocysteine in proteins: Properties and biotechnological use. *Biochimica et Biophysica Acta, 1726*, 1–13.
3. 3. Brown, K. M., & Arthur, J. R., (2001). Selenium, selenoproteins, and human health: A review. *Public Health Nutrition, 4*(2B), 593–599.
4. Jun, L., & Arne, H., (2009). Selenoproteins. *The Journal of Biological Chemistry, 284*(2), 723–727.
5. http://www.ddbj.nig.ac.jp/sub/ref3-e.html (accessed on 6 July 2020). DNA Data Bank of Japan.
6. Marott, S. F., Donald, D. V. S., & David, R. C., (1959). *The Source and State of the Hydroxylysine of Collagen*. II. Failure of free hydroxylysine to serve as a source of the hydroxylysine or lysine of collagen. http://www.jbc.org/content/234/4/918.full.pdf (accessed on 29 June 2020).
7. http://www.hmdb.ca/metabolites/HMDB02038 (*Methyl Lysine*) (accessed on 29 June 2020).
8. http://www.uscnk.com/directory/Gamma-carboxyglutamic-Acid(Gla)-3408.htm (accessed on 29 June 2020).
9. http://www.hmdb.ca/metabolites/HMDB00572 (*Desmosine*) (accessed on 29 June 2020).
10. http://www.hmdb.ca/metabolites/hmdb00214 (Ornithine) (accessed on 29 June 2020).
11. http://www.hmdb.ca/metabolites/HMDB13716 (Norvaline) (accessed on 29 June 2020).
12. ttp://www.elsevierhealth.co.uk/media/us/samplechapters/9780323053716/Chapter%2002.pdf (accessed on 29 June 2020).
13. http://www.aminoacidsguide.com/Trp.html (accessed on 29 June 2020).
14. http://www.aminoacidsguide.com/Pro.html (accessed on 29 June 2020).
15. http://www.nlm.nih.gov/medlineplus/ency/article/002222.htm (accessed on 29 June 2020).
16. Berg, J. M., Tymoczko, J. L., & Stryer, L., (2002). *Biochemistry* (5th edn.) New York: W H Freeman. Section number 23.5, Carbon atoms of degraded amino acids emerge as major metabolic intermediates. Available from: http://www.ncbi.nlm.nih.gov/books/NBK22453/ (accessed on 29 June 2020).
17. Carl, M. S., Philip, E. H., & Richard, P. G., (1951). Occurrence of D-amino acids in some natural materials. *J. Biol. Chem., 190*, 705–710.
18. Mendel, F., (1999). Chemistry, nutrition, and microbiology of D-amino acids. *J. Agric. Food Chem., 47*, 3457–3479.

19. Ryuichi, K., & Yosihiro, Y., (1992). D-Amino-acid oxidase and its physiological function. *International journal of biochemistry, 24(4)*, 519-524.

20. Jumpei, S., Masataka, S., Nobuaki, I., & Sadakazu, A., (2014). Activity of D-amino acid oxidase is widespread in the human central nervous system. *Front. Synaptic Neurosci.*

21. http://global.oup.com/us/companion.websites/fdscontent/uscompanion/us/static/comp anion.websites/9780199730841/McKee_Chapter5_Sample.pdf (accessed on 29 June 2020).

22. Voet, D., & Voet, J. G., (2011). *Biochemistry* (4th edn.). Wiley Publishers.

23. http://www.eu.elsevierhealth.com/media/us/samplechapters/9780323053716/ Chapter%2002.pdf (accessed on 29 June 2020).

24. Aragon, C., & Lopez, C. B., (2003). Structure, function, and regulation of glycine neurotransporters. *Eur. J. Pharmacol., 479*(1–3), 249–262.

25. http://www.altmedrev.com/archive/publications/12/2/169.pdf (2007). *Alternative Medicine Review* (Vol. 12, No. 2). Lysine Monograph.

26. Sidney, M. M. J., (2006). Arginine: Beyond protein. *Am. J. Clin. Nutr., 83*, 508S–512S.

27. http://www.cancer.org/cancer/cancercauses/othercarcinogens/athome/aspartame (accessed on 29 June 2020).

28. Butchko, H. H., Stargel, W. W., Comer, C. P., Mayhew, D. A., Benninger, C., Blackburn, G. L., De Sonneville, L. M., et al., (2002). Aspartame: Review of safety. *Regul. Toxicol. Pharmacol., 35*(2 Pt 2), S1–93.

29. Kulkarni, C., Kulkarni, K. S., & Hamsa, B. R., (2005). L-Glutamic acid and glutamine: Exciting molecules of clinical interest. *Indian J. Pharmacol., 37*(3), 148–154.

30. Pravina, P., (2013). Cysteine-master antioxidant. *IJPCBS, 3*(1), 143–149.

31. http://www.ncbi.nlm.nih.gov/books/NBK27940/ (accessed on 29 June 2020).

32. Dawn, M. R., Michael, A. D., Charles, W. M., Ashley, A., Hill-Kapturczak, N., & Donald, M. D., (2009). L-Tryptophan: Basic metabolic functions, behavioral research, and therapeutic indications. *International Journal of Tryptophan Research, 2*, 45–60.

33. John, D. F., & Madelyn, H. F., (2007). Tyrosine, phenylalanine, and catecholamine synthesis and function in the brain. *J. Nutr., 137*, 1539S–1547S.

34. ehrs.org.uk (accessed on 29 June 2020).

35. Murray, R. K., Granner, D. K., & Rodwell, (2006). *Harper's Illustrated Chemistry* (27th edn.). McGraw Hill Publishers.

36. Gilbert, H. F., (2010). *Peptide Bonds, Disulfide Bonds and Properties of Small Peptides.* In ELS. John Wiley & Sons Ltd, Chichester. http://www.els.net (accessed on 29 June 2020). [doi: 10.1002/9780470015902.a0001328.pub2].

37. (1984). Nomenclature and symbolism for amino acids and peptides. *Pure and Appl. Chem., 56*(5), 595–624.

38. Berg, J. M., Tymoczko, J. L., & Stryer, L., (2002). *Biochemistry* (5th edn.) New York: W H Freeman. Section number 3.2, *Primary Structure: Amino Acids are Linked by Peptide Bonds to Form Polypeptide Chains.* Available from: http://www.ncbi.nlm.nih.gov/ books/NBK22364/ (accessed on 29 June 2020).

39. David, A., & Luis, R., (1998). Animal antimicrobial peptides: An Overview. *Biopolymers (Peptide Science), 47*, 415–433.

40. Janice, C. F., (1997). *Opioid Peptides, 21*(2), 132–136.

41. Anupama, K., Sreemantula, S., & Shaik, R., (2009). Endogenous opioids: Their physiological role and receptors. *Global Journal of Pharmacology, 3*(3), 149–153.

42. Vachkova, E. G., & Bivolarski, B. L., (2007). Origin, structure, and physiological role of the epidermal growth factor: A review. *Bulgarian Journal of Veterinary Medicine, 10*(4), 223–233.

43. Luigi, A., Maria, L. R., Patrizia, B., & Luigi, M., (2012). Nerve growth factor: From the early discoveries to the potential clinical use. *Journal of Translational Medicine, 10*, 239.

44. Sang, H. J., (2012). Cyclic peptides as therapeutic agents and biochemical tool. *Biomol. Ther., 20*(1), 19–26.

45. Hubbard, R. E., & Kamran, H. M., (2010). *Hydrogen Bonds in Proteins: Role and Strength*. In ELS John Wiley & Sons Ltd, Chichester. http://www.els.net (accessed on 29 June 2020). doi: 10.1002/9780470015902.a0003011.pub2.

46. Matthews, B. W., (2001). *Hydrophobic Interactions in Proteins*. eLS.

47. http://www.nature.com/scitable/topicpage/protein-structure-14122136 (accessed on 29 June 2020).

48. Zachary, S. H., & Bruce, T., (1994). Do salt bridges stabilize proteins? A continuum electrostatic analysis. *Protein Science, 3*, 211–226.

49. Alice, Q. Z., Corey, S. O., & Lynne, R. (2011). Revisiting the Ramachandran plot from a new angle. *Protein Science, 20*, 1166–1171.

50. Scott, A. H., & Andrew, K. P., (2010). A fresh look at the Ramachandran plot and the occurrence of standard structures in proteins. *Biomol. Concepts, 1*(3/4), 271–283.

51. Bosco, K. H., & Robert, B., (2005). The Ramachandran plots of glycine and pre-proline. *BMC Structural Biology, 5*, 14. doi: 10.1186/1472-6807-5-14.

52. http://www.proteinstructures.com/Structure/Structure/Ramachandran-plot.html (accessed on 29 June 2020).

53. Berg, J. M., Tymoczko, J. L., & Stryer, L., (2002). *Biochemistry* (5th edn.). New York: W H Freeman. Section 3.2, *Primary Structure: Amino Acids are Linked by Peptide Bonds to Form Polypeptide Chains*. Available from: http://www.ncbi.nlm.nih.gov/books/NBK22364/ (accessed on 29 June 2020).

54. David, E., (2003). *The Discovery of the α-Helix and β-Sheet, the Principal Structural Features of Proteins* (Vol. 100, No. 20, pp. 11207–11210). Howard Hughes Medical Institute and University of California-Department of Energy Institute of Genomics and Proteomics, University of California, Los Angeles, CA 90095-1570; PNAS September 30.

55. Linus, P., Robert, B. C., & Branson, H. R., (1951). *The Structure of Proteins: Two Hydrogen-Bonded Helical Configurations of the Polypeptide Chain*. Gates and Crellin Laboratories of Chemistry, California Institute of Technology, Pasadena, California Communicated. Chemistry: Pauling, Corey, Branson. *Proc. N. A. S.*

56. Linus, P., & Robert, B. C., (1951). *The Pleated Sheet: A New Layer Configuration of Polypeptide Chains* (Vol. 37). Gates and Crellin Laboratories of Chemistry, California Institute of Technology, Pasadena, California; Communicated, Chemistry: Pauling and Corey.

57. Leon, N. H., (1972). Structural aspects of keratin fibers. *J. Soc. Cosmet. Chem., 23*, 427–445.

58. Carola, D., Gareth, C., & Patrick, A., (1994). Structural characteristics and stabilizing principles of bent β-strands in protein tertiary architecture. *Protein Science, 3*, 876–882.

59. Ananthanarayanan, V. S., Attah-Poku, S. K., Mukkamala, P. L., & Rehse, P. H., (1985). Structural and functional importance of the β-turn in proteins. *Proc. Int. Symp. Biomol. Struct. Interactions, Suppl. J. Biosci., 8*(1/2), 209–221.

60. Nick, P. J. R., Patrick, J. F., & George, D. R., (2005). Hydrogen-bonded turns in proteins: The case for a recount. *Protein Science, 14*, 2910–2914.

61. Jacquelyn, S. F., (1995). Omega loops: Nonregular secondary structures significant in protein function and stability. *FASEB, J., 9*, 708–717.

62. Arnab, B., & Manju, B., (2005). Collagen structure: The madras triple helix and the current scenario. *IUBMB Life, 57*(3), 161–172.

63. Berg, J. M., Tymoczko, J. L., & Stryer, L., (2002). *Biochemistry* (5th edn.) New York: W H Freeman; Section 3.4. *Tertiary Structure: Water-Soluble Proteins Fold Into Compact Structures with Nonpolar Cores.* Available from: http://www.ncbi.nlm.nih.gov/books/NBK22375/ (accessed on 29 June 2020).

64. http://www.rcsb.org/pdb/101/motm.do?momID=36 (accessed on 29 June 2020).

65. Margoliash, E., (1963). Primary structure and evolution of cytochrome C. *Proc. N. A. S., 50*, 672–679.

66. http://www.uniprot.org/uniprot/P99999 (accessed on 29 June 2020).

67. George, A. O., & Daniel, J. G., (2004). Myoglobin: An essential hemoprotein in striated muscle. *The Journal of Experimental Biology, 207*, 3441–3446.

68. Berg, J. M., Tymoczko, J. L., & Stryer, L., (2002). *Biochemistry* (5th edn.) New York: W H Freeman, Section 3.5, *Quaternary Structure: Polypeptide Chains Can Assemble Into Multisubunit Structures.* Available from: http://www.ncbi.nlm.nih.gov/books/NBK22550/ (accessed on 29 June 2020).

69. Robert, J. H., William, K., Guido, G., & Lyman, C. C., (1962). The structure of human hemoglobin I. The separation of the α and β chains and their amino acid composition. *TheJournal of Biological Chemistry, 237*(5), 1549–1554.

70. Alan, N. S., (2008). Hemoglobin research and the origins of molecular medicine. *Blood, 112*(10), 3927–3938.

71. Marengo-Rowe, A. J., (2006). Structure-function relations of human hemoglobin's. *Proc. (Bayl. Univ. Med. Cent), 19*, 239–245.

72. White, J. C., & Beaven, G. H., (1954). Review of the varieties of human hemoglobin in health and disease. *J. Clin. Path., 7*, 175.

73. http://www.nlm.nih.gov/medlineplus/ency/article/000562.htm (accessed on 29 June 2020).

74. Hue, S. C., & Ken, A. D., (1990). Origins of structure in globular proteins. *Proc. Nati. Acad. Sci. USA, 87*, 6388–6392.

75. Lisa, D. M., & Fred, W. K., (2013). Molecular assembly and mechanical properties of the extracellular matrix: A fibrous protein perspective. *Biochimica et Biophysica Acta, 1832*, 866–875.

76. Rod, B., (2007). The protamine family of sperm nuclear proteins. *Genome Biology, 8*, 227. doi: 10.1186/gb-2007-8-9-227.

77. Rafael, O., (2006). Protamines and male infertility. *Human Reproduction Update, 12*(4), 417–435.

78. Snustad, D. P., Simmons, M. J., (2011). Genetics. Philadelphia: John Wiley & Sons.

79. Ralph, K. L., (1927). *A Comparative Study of the Glutelins of the Cereal Grains, 35*(12), 1091–1120.

80. http://www.fao.org/docrep/x5738e/x5738e04.htm (accessed on 29 June 2020).

81. *Metalloproteins: Chromatography.* E. Parisi, in Encyclopedia of Separation Science, Academic Press, London 2000, pp. 3380 – 3386..

82. 82. Achyuthan, K. E., & Ramachandran, L. K., (1983). Non-identity of reaction center's for pyrophosphatase and toxic actions of cardiotoxin II: The status of cardiotoxin II as a metalloprotein. *J. Biosci., 5*(1), 1–6.

83. 83. Jacqueline, M. F., Anna, M. D., Charlotte, G., & Jean, R., (1950). The distribution of the chromoproteins, hemoglobin, myoglobin, and cytochrome C, in the tissues of different species, and the relationship of the total content of each chromoprotein to body mass. David L. Drabkin with the technical assistance of Priscilla Fourier. *J. Biol. Chem., 182*, 317–334.

84. Schmid, K. A. I. U., (1971). Characterization and structure of plasma glycoproteins. Pure and applied chemistry. *Chimie pure et appliquee, 27(4)*, 591–596.

85. Frédéric, D., & Eric, C., (2006). Phosphoprotein analysis: From proteins to proteomes. *Proteome Science, 4*, 15. doi: 10.1186/1477-5956-4-15.

86. Nestler, E. J., & Greengard, P., (1999). Neuronal phosphoproteins. In: Siegel, G. J., Agranoff, B. W., Albers, R. W., et al., (eds.), *Basic Neurochemistry: Molecular, Cellular and Medical Aspects* (6th edn.) Philadelphia: Lippincott-Raven. Available from: http://www.ncbi.nlm.nih.gov/books/NBK28010/ (accessed on 29 June 2020).

87. Cox, R. A., & García-Palmieri, M. R., (1990). Cholesterol, triglycerides, and associated lipoproteins. In: Walker, H. K., Hall, W. D., & Hurst, J. W., (eds.), *Clinical Methods: The History, Physical, and Laboratory Examinations* (3rd edn.). Boston: Butterworths: Chapter 31. Available from: http://www.ncbi.nlm.nih.gov/books/NBK351/ (accessed on 29 June 2020).

88. Chittenden, R. H., & Hartwell, J. A., (1887). The relative formation of proteoses and peptones in gastric digestion. *Jahresberichtf and IrThierchemieffur*, 264.

89. Alberts, B., Johnson, A., Lewis, J., et al., (2002). *Molecular Biology of the Cell* (4th edn.). New York: Garland Science. Protein Function. Available from: http://www.ncbi.nlm.nih.gov/books/NBK26911/ (accessed on 29 June 2020).

90. http://www.nature.com/scitable/topicpage/protein-function-14123348 (accessed on 29 June 2020).

91. Alan, N. S., (2008). Hemoglobin research and the origins of molecular medicine. *Blood, 112*(10), 3927–3938.

92. Scientia, S., (1973). Studies on the structure-function relationships of insulin I: The relationship of the C-terminal peptide sequence of B-chain to the activity of insulin. *The Shanghai Insulin research Group, 14*(1), 61–70.

93. Keith, D. K., & Zoran, O., (2001). The protein trinity-linking function and disorder. *Nature Biotechnology, 19*, 805–806.

94. Janet, M. T., Annabel, E. T., Duncan, M., Neera, B., & Christine, A. O., (2000). From structure to function: Approaches and limitations. *Nature Structural Biology*, *Structural Genomics Supplement*, 991–994.

95. Alberts, B., Johnson, A., Lewis, J., et al., (2002). *Molecular Biology of the Cell* (4th edn.). New York: Garland Science.

96. Berg, J. M., Tymoczko, J. L., & Stryer, L., (2002). *Biochemistry* (5th edn.). New York: W H Freeman, Section 4.5. *Three-Dimensional Protein Structure can be Determined by NMR Spectroscopy and X-Ray Crystallography*. Available from: http://www.ncbi.nlm.nih.gov/books/NBK22393/ (accessed on 29 June 2020).

97. Alberts, B., Johnson, A., Lewis, J., et al., (2002). *Molecular Biology of the Cell* (4th edn.). New York: Garland Science, Analyzing protein structure and function. Available from: http://www.ncbi.nlm.nih.gov/books/NBK26820/Physical Properties of Proteins (accessed on 29 June 2020).

98. Lisa, D. M., & Fred, W. K., (2013). Molecular assembly and mechanical properties of the extracellular matrix: A fibrous protein perspective. *Biochimica et Biophysica Acta, 1832*, 866–875.

99. Saul, R. T., Martin, S. J., & Nick, P. C., (2008). Measuring and increasing protein solubility. *Journal of Pharmaceutical Sciences, 97*, 4155–4166.

100. Wade L. G., Simek J. W., Singh M. S., (2019). Organic Chemistry. New Delhi: Pearson India.

101. Inglis, A. S., Nicholls, P. W., & Roxburgh, O. M., (1971). Hydrolysis of the peptide bond and amino acid modification with hydriodic acid. *Aust. J. biol. Sci., 24*, 1235–1240.

102. Snustad, D. P., (2019). Principles of Genetics. Philadelphia: John Wiley.

103. https://www.bdbiosciences.com/documents/Hydrolysis_to_Hydrolysate.pdf (accessed on 29 June 2020).

104. http://www.worthington-biochem.com/try/(accessed on 29 June 2020).

105. Zahid, H. C., Arif, M., Muhammad, A. A., & Claudiu, T. S., (2006). Metal-based antibacterial and antifungal agents: Synthesis, characterization, and biological evaluation of Co(II), Cu(II), Ni(II), and Zn(II) complexes with amino acid-derived compounds. *Bioinorg. Chem. Appl.*, 83131. PMCID: PMC1800917.

106. Belitz, H. D., Grosch, W., Schieberle, P., (2004). Amino Acids, Peptides, Proteins. In: Food Chemistry. Springer, Berlin, Heidelberg.

107. Walker, J. M., (1984). The dansyl method for identifying N-terminal amino acids. *Methods Mol. Biol., 1*, 203–212. doi: 10.1385/0-89603-062-8, 203.

108. Neuberger, A., (1944). The reaction between histidine and formnaldehyde. *Biochem. J., 38*(4), 309–314.

109. http://books.google.co.in/books?id=BVpDI7n2M9gC&pg=PA56&lpg=PA56&dq=sore nsen+titration&source=bl&ots=gYDwOcfJlC&sig=jQWHLfDSBl228P5LXDAU1RiJ OrU&hl=en&sa=X&ei=6ghfVOGOL4OwuASo8oLIBw&ved=0CGEQ6AEwCTgK#v =onepage& q=sorensen%20titration& f=false (accessed on 29 June 2020).

110. http://books.google.co.in/books?id=I3gC0bX_IKAC&pg=PA81&lpg=PA81&dq=rea ction+of+nh2+group+of+amino+acids&source=bl&ots=CpxNGHhJ4a&sig=Bq1mA ZwRZYyKkj_SVaoncj1KWPQ&hl=en&sa=X&ei=zQ9fVJb5BZSQuATN3ICoCA& ved=0CEEQ6AEwBw#v=onepage&q=reaction%20of%20nh2%20group%20of%20 amino%20acids&f=false (accessed on 29 June 2020).

111. Donald, D. V. S., (1929). Manometric determination of primary amino nitrogen and its application to blood analysis. *J. Biol. Chem., 83*, 425–447.

112. http://books.google.co.in/books?id=Sj6Xrc78LKUC&pg=PA25&lpg=PA25&dq=van +slyke+reaction+1911&source=bl&ots=xIL1f7WywU&sig=olTufQ9bhivkFckPR1C BR3zTYow&hl=en&sa=X&ei=GhRfVKeIHY61uAThhoDIDg&ved=0CDsQ6AEwB Q#v=onepage&q=van%20slyke%20reaction%201911&f=false(accessed on 29 June 2020).

113. Addison, A., (2004). The monosodium glutamate story: The commercial production of MSG and other amino acids. *J. Chem. Edu., 81*(3), 347–355.

114. (1968). *Can Med Assoc J., 99*(24), 1206–1207. PMCID: PMC1945602; Chinese restaurant syndrome.

115. http://www.nlm.nih.gov/medlineplus/ency/article/001126.htm (accessed on 29 June 2020).

116. Friedman, M., (2004). Applications of the ninhydrin reaction for analysis of amino acids, peptides, and proteins to agricultural and biomedical sciences. *J. Agric. Food Chem., 52*(3), 385–406.

117. Lapidot, Y., & Katchalski, E., (1968). Reaction of phosgene with serine and ethanolamine. *Isr. J. Chem., 6*, 147–150. doi: 10.1002/ijch.196800022.

118. Busher, J. T., (1990). Serum albumin and globulin. In: Walker, H. K., Hall, W. D., & Hurst, J. W., (eds.), *Clinical Methods: The History, Physical, and Laboratory Examinations* (3rd edn.). Boston: Butterworths; Chapter number 101. Available from: http://www.ncbi.nlm.nih.gov/books/NBK204/ (accessed on 29 June 2020).

119. Harisha, (2008). *Biotechnology Procedures and Experiments Handbook.* Firewell Media Publishers.

120. Tzong-Shi, L., Szu-Yu, Y., Kenneth, L., Roderick, V. J., & Li-Li, H., (2010). Interpretation of biological and mechanical variations between the Lowry versus Bradford method for protein quantification. *N. Am. J. Med. Sci., 2*(7), 325–328. doi: 10.4297/najms.2010.2325; PMCID: PMC3341640.

121. Tao, H., Mian, L., & Bo, H., (2010). Competitive binding to cuprous ions of protein and BCA in the bicinchoninic acid protein assay. *The Open Biomedical Engineering Journal, 4*, 271–278.

122. Lodish, H., Berk, A., Zipursky, S. L., et al., (2000). *Molecular Cell Biology* (4th edn.) New York: W. H. Freeman, Section 17.4. Translocation of secretory proteins across the ER membrane. Available from: http://www.ncbi.nlm.nih.gov/books/NBK21532/ (accessed on 29 June 2020).

123. Cooper, G. M., (2000). *The Cell: A Molecular Approach* (2nd edn.) Sunderland (MA): Sinauer Associates. The Endoplasmic Reticulum. Available from: http://www.ncbi.nlm.nih.gov/books/NBK9889/ (accessed on 29 June 2020).

124. Lodish, H., Berk, A., Zipursky, S. L., et al., (2000). *Molecular Cell Biology* (4th edn.) New York: W. H. Freeman, Section 17.3, Overview of the secretory pathway. Available from: http://www.ncbi.nlm.nih.gov/books/NBK21471/ (accessed on 29 June 2020).

125. Cooper, G. M., (2000). *The Cell: A Molecular Approach* (2nd edn.). Sunderland (MA): Sinauer Associates. The Golgi apparatus. Available from: http://www.ncbi.nlm.nih.gov/books/NBK9838/ (accessed on 29 June 2020).

126. http://books.google.co.in/books?id=R0W1ErpsQpkC&pg=PA187&lpg=PA187&dq=topological+modifications+of+peptide&source=bl&ots=oCMpi1_Zpr&sig=HN_1lMvj5BI5wh9fdzS0jKejYSY&hl=en&sa=X&ei=r3FgVOTXDNiOuAS-v4DABA&sqi=2&ved=0CD4Q6AEwBA#v=onepage&q=topological%20modifications%20of%20peptide&f=false (accessed on 29 June 2020).

127. http://books.google.co.in/books?id=XM2_SX953W0C&pg=PA97&lpg=PA97&dq=pseudo+peptide+bond+in+proteins&source=bl&ots=vDKKRkYzdq&sig=QIRkHrJYVZfPmKPQYfleqglhiqY&hl=en&sa=X&ei=xp1gVMeLIcSRuAT_g4D4BA&ved=0CFMQ6AEwDA#v=onepage&q=pseudo%20peptide%20bond%20in%20proteins&f=false(accessed on 29 June 2020).

128. http://www.nlm.nih.gov/medlineplus/ency/article/002380.htm(accessed on 29 June 2020).

129. Niscair Tutorial. http://nsdl.niscair.res.in/jspui/bitstream/123456789/759/1/revised%20Protein%20and%20polypeptide%20hormones.pdf (accessed on 29 June 2020).

130. Shalev, E., & Leung, P. C., (2003). Gonadotropin-releasing hormone and reproductive medicine. *J. Obstet. Gynaecol. Can., 25*(2), 98–113.

131. Nussey, S., & Whitehead, S., (2001). *Endocrinology: An Integrated Approach*. Oxford: BIOS Scientific Publishers, Chapter 7, The pituitary gland. Available from: http://www.ncbi.nlm.nih.gov/books/NBK27/ (accessed on 29 June 2020).

132. Carter-Su, C., Schwartz, J., & Smit, L. S., (1996). Molecular mechanism of growth hormone action. *Annu. Rev. Physiol., 58*, 187–207.

133. Takeda, A., Cooper, K., Bird, A., Baxter, L., Frampton, G. K., Gospodarevskaya, E., Welch, K., & Bryant, J., (2010). Recombinant human growth hormone for the treatment of growth disorders in children: A systematic review and economic evaluation. *Health Technol. Assess, 14*(42), 1–209, III-IV. doi: 10.3310/hta14420.

134. Peter, J. T. M. D., William, M. D. M. B., Laurence, K. M. D., Pamela, U. F. M. D., Herman-Bonert, V. M. D., Van, D. L. A. J. M. D., Eleni, V. D. M. D., et al., (2000). Treatment of Acromegaly with the growth hormone-receptor antagonist pegvisomant. *N Engl. J. Med., 342*, 1171–1177. doi: 10.1056/NEJM200004203421604.

135. Nussey, S., & Whitehead, S., (2001). *Endocrinology: An Integrated Approach*. Oxford: BIOS Scientific Publishers. Chapter 6, The gonad. Available from: http://www.ncbi.nlm.nih.gov/books/NBK29/ (accessed on 29 June 2020).

136. Goa, K. L., & Wagstaff, A. J., (1998). Follitropin alpha in infertility: A review. *Bio. Drugs, 9*(3), 235–260.

137. Nussey, S., & Whitehead, S., (2001). *Endocrinology: An Integrated Approach*. Oxford: BIOS Scientific Publishers. Chapter 4, The adrenal gland. Available from: http://www.ncbi.nlm.nih.gov/books/NBK26/ (accessed on 29 June 2020).

138. http://www.nlm.nih.gov/medlineplus/ency/article/003696.htm (accessed on 29 June 2020).

139. Jacob, G., Joseph, P. C., & Karl, M., (2009). H.P. Acthar, gel and cosyntropin review clinical and financial implications. *P& T, 34*(5), 250–257.

140. Nussey, S., & Whitehead, S., (2001). *Endocrinology: An Integrated Approach*. Oxford: BIOS Scientific Publishers. Chapter 7, The pituitary gland. Available from: http://www.ncbi.nlm.nih.gov/books/NBK27/ (accessed on 29 June 2020).

141. http://pubs.acs.org/cen/coverstory/83/8325/8325oxytocin.html (accessed on 29 June 2020).

142. 142. Gomez, D. E., Ripoll, G. V., Girón, S., & Alonso, D. F., (2006). Desmopressin and other synthetic vasopressin analogs in cancer treatment. *Bull Cancer, 93*(2), E7–12.

143. Nussey, S., & Whitehead, S., (2001). *Endocrinology: An Integrated Approach*. Oxford: BIOS Scientific Publishers. Chapter 5, The parathyroid glands and vitamin D. Available from: http://www.ncbi.nlm.nih.gov/books/NBK24/ (accessed on 29 June 2020).

144. Michael, M. M. D., Bart, L. C., Tamara, V., Maria, L. B., Lakshminarayan, R., William, D. F., Peter, L., et al., (2013). Efficacy and safety of recombinant human parathyroid hormone (1–84) in hypoparathyroidism (REPLACE): A double-blind, placebo-controlled, randomized, phase 3 study. *The Lancet Diabetes and Endocrinology, 1*(4), 275–283.

145. Nussey, S., & Whitehead, S., (2001). *Endocrinology: An Integrated Approach*. Oxford: BIOS Scientific Publishers. Chapter 2, The endocrine pancreas. Available from: http://www.ncbi.nlm.nih.gov/books/NBK30/ (accessed on 29 June 2020).

146. Vasiliki, V., (2010). Therapeutics of diabetes mellitus: Focus on insulin analogs and insulin pumps. *Experimental Diabetes Research, 2010*, 14. Article ID 178372. doi: 10.1155/2010/178372.

147. Jessy, S., & Patole, V., (2008). Protein and peptide drug delivery: Oral approaches. *Indian J. Pharm. Sci., 70*(3), 269–277. doi: 10.4103/0250-474X.42967. PMCID: PMC2792531.

148. Ratnaparkhi, M. P., Chaudhari, S. P., & Pandya, V. A., (2011). Peptides and proteins in pharmaceuticals. *Int. J. Curr. Pharm. Res., 3*(2), 1–9.

149. https://www.inkling.com/read/lippincotts-pharmacology-harvey-champe-5th/chapter-1/ii--routes-of-drug (accessed on 29 June 2020).

150. Pooja, J., Paresh, M., & Sandip, P., (2012). Pharmaceutical approaches related to systemic delivery of protein and peptide drugs: An overview. *International Journal of Pharmaceutical Sciences Review and Research, 12*(1), 42–52. Article 7.

151. http://books.google.co.in/books?id=8BaT2dIX86AC&pg=PA3&lpg=PA3&dq=methods+of+peptide+synthesis+review&source=bl&ots=BvSKFGTcTN&sig=0LHE7fcQ1zo7VTCab_eYhu5XBok&hl=en&sa=X&ei=R2FjVPH2OKO4mAWF1oLoCw&ved=0CBsQ6AEwADgU#v=onepage&q=methods%20of%20peptide%20synthesis%20review&f=false (accessed on 29 June 2020).

152. Saranya, C., Pavla, S., & Istvan, T., (2013). Chemical methods for peptide and protein production. *Molecules, 18*, 4373–4388. doi: 10.3390/molecules18044373.

153. http://books.google.co.in/books?id=ODjiBAAAQBAJ&pg=PA316&lpg=PA316&dq=peptides+and+proteins+mediators&source=bl&ots=WIacQ8De2V&sig=ZfuBZBlr7Uc4J1nRRiY0PH_3xvo&hl=en&sa=X&ei=l3ljVMyHDMehugTQrYJw&ved=0CDsQ6AEwBw#v=onepage&q=peptides%20and%20proteins%20mediators&f=false (accessed on 29 June 2020).

CHAPTER 3

Biotechnological Potential of Hydrogen-Oxidizing Bacteria

TATIANA GRIGORIEVNA VOLOVA,[1,2] EVGENIY GENNADIEVICH KISELEV,[1,2] and EKATERINA IGOREVNA SHISHATSKAYA[1,2]

[1]Siberian Federal University, 79 Svobodnyi Av., Krasnoyarsk – 660041, Russia

[2]Institute of Biophysics SB RAS, Federal Research Center "Krasnoyarsk Science Center SB RAS," 50/50 Akademgorodok, Krasnoyarsk – 660036, Russia, E-mail: volova45@mail.ru (T. G. Volova)

3.1 INTRODUCTION

Growing rates of consumption of food substances and energy resources, population increase, and human-induced impacts have reached the limit beyond which uncoordinated economic activity can lead to irreversible changes in the biosphere. The only way to achieve harmonic coexistence between the growing population of our planet and the biosphere is to pursue coordinated development of science and technology. Development of novel, closed-loop, technologies for complete treatment of natural resources and production of environmentally friendly energy sources and materials that can be involved in the biospheric cycling are the objectives consistent with the conception of the "agenda of the 21st century" adopted at the special UN session on the environment and development.

Achievements of modern biology, which have led to the formation of physicochemical biology and a number of new branches of biotechnology, exert a qualitative effect on various spheres of human activity. The demand for them is growing as is their ability to solve the key issues of the 21st century. The diversity of forms of living matter and the new knowledge on the physics and chemistry of living systems provide a basis for constructing biological systems varied in complexity and organization for the synthesis of a wide spectrum of macromolecules.

Among efficient producers of valuable biotechnological materials are chemolithotrophic hydrogen-oxidizing microorganisms, which use hydrogen as the main growth substrate. Hydrogen bacteria proper (bacteria oxidizing hydrogen, or "detonating gas" bacteria) were almost simultaneously isolated in the early 20th century by Kaserer [1] and Lebedeff [2, 3], using Vinogradskii's elective method. The organisms became known as hydrogen bacteria proper or *Knallgasbakterien* (detonating gas bacteria).

One cannot overestimate the scientific achievements of S. N. Vinogradskii, a great Russian scientist, whose name is directly related to the discovery of chemosynthesis as a new type of metabolism. That, together with anaerobism discovered by L. Pasteur, provided insight into exceptional physiological specialization of microorganisms, served as the basis for general microbiology and modern comparative biochemistry, and significantly influenced very many concepts of modern biology.

The successful studies of hydrogen-oxidizing bacteria in the 1920s–1930s demonstrated the diversity of these organisms and their specific metabolism. The general interest in chemoautotrophy that increased in the early 1950s was particularly focused on hydrogen bacteria, which were used as model organisms in the studies addressing not only chemoautotrophy but also general biochemical mechanisms of microbial metabolism. Studies on hydrogen biosynthesis were inspired and organized by Academician, Zavarzin [4]. Numerous hydrogen-oxidizing microorganisms from various geographic areas were isolated, systematized, studied, and offered to other research institutes. Hydrogen bacteria were actively and successfully studied in Germany by Professor H. H. Schlegel and his coworkers and in the USA, Japan, and other countries.

Practical interest in hydrogen bacteria was related to their potential use as a regenerative component in closed biotechnological life support systems (LSSs). Biosynthesis of hydrogen bacteria in combination with electrolysis was proposed as a solution to the main issues of human life support during space flights: oxygen generation and carbon dioxide consumption, treatment of the water to remove metabolites, and production of the protein part of the human diet.

More recently, hydrogen bacteria, which grow much faster than other chemoautotrophic organisms, attracted the attention of researchers as a potential source of feed, or even food, protein. Their metabolism and growth were extensively studied in the USA, the FRG, the USSR, and Japan. The late 1980s and the early 1990s saw an upsurge of interest in hydrogen bacteria as very promising producers of polyhydroxyalkanoates (PHAs) (polyesters of alkanoic acids)-polymers similar to polypropylene but degradable in the natural environment.

The study of hydrogen-oxidizing bacteria at the Institute of Biophysics of the SB RAS was started in the late 60s to early 70s of the last century. Later, in collaboration with the Siberian Federal University, a scientific team was formed with extensive experience in the field of hydrogen biosynthesis; the path has gone from culturing bacteria in small laboratory apparatuses to pilot plants. Our studies of hydrogen-oxidizing bacteria included several areas:

1. **The Hydrogen Bacteria-Regenerative Link of Human Life Support Systems (LSSs):** It has been shown that a culture of hydrogen together with water electrolysis, can solve the main tasks of human LSSs: oxygen supply, assimilation of released CO_2, utilization of human wastes and water treatment, and synthesis of protein-rich biomass.

2. **Biomass of Hydrogen Bacteria (BHB)-Potential Protein Source:** A technology has been developed for the synthesis of protein biomass on hydrogen; 10 tons of biomass was studied in rations of tiptsy, farm animals, and fur-bearing animals. The possibility of replacing 35–50% of proteins in BHB feeds without a decrease in animal productivity was shown.

3. **Hydrogen Bacteria and Synthesis of Biodegradable Polymers:** The ability of hydrogen-oxidizing bacteria to synthesize PHA under autotrophic and heterotrophic conditions makes them good candidates for commercial production of PHAs. Researchers of the Institute of Biophysics and SFU tested different modes of cultivation of hydrogen bacteria on the different substrate (electrolytic hydrogen; synthesis gas; sugars, organic, and fatty acids, vegetable oil [32, 33].

This chapter reviews these results.

3.2 RESULTS AND DISCUSSION

Hydrogen-oxidizing bacteria are characterized by high growth rates, the absence of intermediate carbon-containing metabolic products, metabolism variability, and a broad organotrophic potential. Hydrogen-oxidizing bacteria synthesizes high yields of protein under conditions of unlimited growth, hydrogen-oxidizing bacteria synthesizes high yields of protein; under unbalanced growth conditions, the cells accumulate PHA.

3.2.1   THE HYDROGEN BACTERIA-REGENERATIVE LINK OF HUMAN LIFE SUPPORT SYSTEMS (LSSS)

Practical uses of hydrogen bacteria are not limited to the production of single–Cell protein or other valuable metabolic products. They can also serve as a potential component of human LSSs. Isolated from the Earth's atmosphere (in spaceships or submarines), under sunlight deficiency, hydrogen bacteria, together with water electrolysis, can solve the main tasks of human life support: oxygen supply, assimilation of released CO_2, utilization of human wastes and water treatment, and synthesis of protein-rich biomass [5–12] (Figure 3.1).

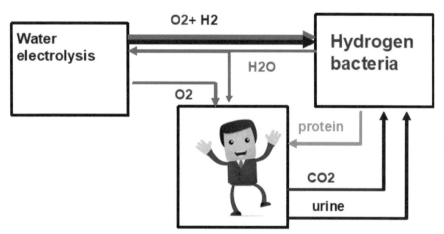

FIGURE 3.1   The hydrogen bacteria, together with water electrolysis, can solve the main tasks of human life support systems: O_2 supply, assimilation of CO_2, utilization of human wastes and water treatment, synthesis of protein-rich biomass.

The most important function of the regeneration component in the life support system is utilization of carbon dioxide expired by humans. Human respiration intensity is characterized by the respiratory quotient (RQ)-[CO_2/O_2]. Idealized gas exchange in humans can be represented as follows:

$$CH_2O + O_2 \rightarrow CO_2 + H_2O$$

The human RQ, in this case, is equal to 1. In reality, it can vary widely (from 0.7 to 0.9), depending on the diet and metabolism intensity. In the LSS with hydrogen bacteria, oxygen regeneration can be performed by electrolysis of water, which accompanies oxidative metabolism of the human and bacteria. Water electrolysis also generates hydrogen – the main growth substrate for bacteria. The gas balance can be achieved in the closed system if RQ of the human is equal to the assimilation quotient (AQ) of the "hydrogen bacteria-electrolysis" system. The studies showed that in combination with water electrolysis ($6H_2O \rightarrow 3O_2 + 6H_2$), the "hydrogen bacteria-electrolysis" system can provide AQ = $[CO_2/O_2]$ = 1. Thus, ideally, the LSS with hydrogen bacteria can be completely gas balanced. The system including the "hydrogen bacteria-electrolysis" unit, intended for one human, will produce 210 l/h of hydrogen. To generate this amount of hydrogen electrolytically, the 252 A current must be passed through the electrolyte for one hour. With voltage 2 V applied to electrodes, the electrical power consumption of the system will amount to 0.5 kW. The powers of the CO_2 concentrator and the bacterial culture compartment are estimated at 0.15 kW and 0.1 kW, respectively. So, the total power of the life support system including hydrogen bacteria and electrolysis, which is intended for one human, will approximately amount to 1 kW (taking into account power measuring instruments) [13].

Another important function of the regeneration component in the life support system is regeneration of water. Culture of hydrogen bacteria in the LSS can utilize human wastes and regenerate water. In the LSS, human wastes can be used as components of the substrate for bacterial culture. This is the way to utilize wastes and regenerate water in the system. Closing of the water loop is a necessary condition for arranging mass exchange in the LSS. It can be attained by recirculating the nutrient medium in the bacterial bioreactor and using human wastes as components of this medium.

Experiments showed that *A. eutrophus* Z1 cells could be grown autotrophically in steady culture for long periods of time on the nutrient medium containing human liquid waste as a nitrogen source, and the gas composition of the culture remained unimpaired [13]. The specific growth rate of bacterial cells utilizing urine components was the highest possible for this strain (up to 0.400 h^{-1}), and the cell culture utilized nitrogenous components efficiently. In addition to urea, bacterial cells utilized uric acid and creatinine. Cells synthesized from this nitrogen substrate had a usual chemical composition and contained large amounts of high-quality proteins. The presence of organic nitrogen in the medium did not produce any adverse effect on the ability of hydrogen bacteria to oxidize molecular hydrogen. The efficiency of utilization of hydrogen oxidation energy (Figure 3.2) by the culture in experiments with human liquid wastes added to the medium was similar to that in experiments with conventional, synthetic, nitrogen sources – NH_4Cl or $CO(NH_2)_2$ [14].

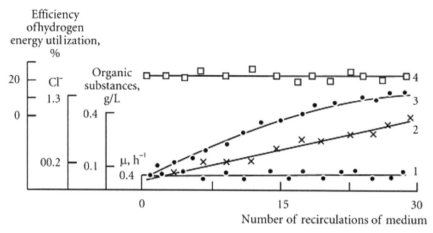

FIGURE 3.2 Parameters of *A. eutrophus* Z1 culture grown on the recirculated medium containing human liquid wastes: (1) specific growth rate (μ, h^{-1}); (2) concentration of organic substances in the culture; (3) concentration of chlorine ions (Cl^-); (4) efficiency of hydrogen energy utilization (%).

Thus, addition of human liquid waste to the recirculated nutrient solution must not lead to accumulation of organic nitrogenous compounds, which would complicate the culturing of hydrogen bacteria. However, human liquid wastes contain not only biogenic elements (nitrogen, phosphorus, sulfur, potassium, magnesium, etc.), but also sodium and chlorine ions, which are not utilized by cells of hydrogen bacteria and, thus, can be accumulated in the recirculated medium.

The culture medium contained increasing amounts of metabolites of hydrogen bacteria and sodium chloride, as a component of urine. The accumulation dynamics of sodium and chlorine ions was examined in experiments with the medium recirculation coefficients 0.5 and 0.9. In both cases, sodium concentration in the culture increased to a certain level and then remained unchanged, producing no effect on specific growth rate of bacterial cells.

Stabilization of sodium concentration in the culture is achieved by continuously removing the element from the medium with harvested biomass. Thus, the amount of the removed sodium is determined by the frequency of biomass harvesting and processes of adsorption on cell surfaces. Chemical tests showed that the amount of sodium removed with cell biomass increased as its concentration in the medium surrounding cells grew higher. Sodium concentration in the medium increased from 0.090 g/L to 1.700 g/L. The amount of sodium in the centrifuged biomass increased from 2.5 mg to 8.5

mg per 1 g of dry biomass, as a result of sodium adsorption on cell surface. At the same time, the intracellular sodium concentration remained unchanged, amounting to 1.4 ± 0.5 mg/g. Thus, knowing the amount of sodium chloride entering the bioreactor with human liquid waste and the amount exiting the bioreactor with harvested biomass, one can calculate the time and value of steady-state NaCl concentration in the culture. It has been experimentally proven that this value must not exceed 7 g/L. Yet, model experiments showed that an increase in NaCl in the culture to 10 g/L did not affect either growth rate or physiological state of *A. eutrophus* Z1 cells. Thus, hydrogen bacteria can be used to utilize human liquid wastes and regenerate water in the LSS.

The total water balance of the system is determined by the following processes: water formation by bacterial cells oxidizing hydrogen, electrolytic water decomposition, and release of metabolic water by humans. The amount of water formed by bacterial cells per 1 g of synthesized biomass can be calculated from the data on the rate of consumption of hydrogen by the cells and elemental composition of biomass. Of the 0.63 g of hydrogen consumed by the culture per 1 g synthesized biomass, about 0.07 g is used to construct cells and the other 0.56 g together with 0.14 g of hydrogen entering the system with urea is oxidized to form water. The amount of the water formed per 1 g synthesized biomass is 5.17 g. The production rate of the culture of hydrogen bacteria necessary to provide gas exchange of one human is 30 g/h, which yields 155.1 g water/h. One human releases about 350 g water/d, or 14.6 g/h. Hence, the water formed in the system amounts to 169.7 g/h. The amount of water decomposed in the electrolytic unit is determined by the amount of hydrogen needed by bacteria-170.1 g/h. Thus, the water balance of the system can reach 0.3%. Moreover, it is not unlikely that due to the flexibility of human metabolisms and that of bacterial cells the water balance could be retained even if the above parameters were altered.

One more important function of the bioregeneration component of the human life support system is production of food substances, primarily the protein components of the diet. Chemically speaking, culture of hydrogen-oxidizing bacteria can fulfill this function. The biomass of *Alcaligenes eutrophus* cells contains large amounts of proteins with a full complement of essential amino acids.

Experiments showed that *Alcaligenes* cells grown in intensive continuous culture on balanced nutrient medium do not synthesize storage macromolecules (such as the polymer of hydroxybutyric acid), which would be unwelcome in this case [15, 16]. Thus, under optimal growth conditions, cells of hydrogen-oxidizing bacteria synthesize mainly nitrogen-containing components (proteins and nucleic acids).

Calculations show that culture of hydrogen bacteria providing gas exchange of one human, included in the life support system, will produce 720 g cell biomass per day. The data on the biochemical and elemental composition of cell biomass and human [12] daily requirements for food components provided a basis for calculating the balance of the mutual supply of each component by the human and the "hydrogen bacteria-electrolysis" system (Table 3.1). The LSS including the "hydrogen bacteria-electrolysis" system that has reached complete equilibriums in the gas and water exchanges can satisfy human requirements for most major elements, proteins, including all essential amino acids, and certain vitamins. However, there will be deficiency of carbohydrates and some mineral elements (sodium, chlorine, iodides, and fluorides). Human liquid wastes contained in the culture media cannot completely satisfy requirements of hydrogen bacteria for the major elements, supplying 20% of nitrogen, 13.4% of phosphorus, 7.5% of magnesium, and 60% of potassium. Therefore, some components will have to be pre-stored in the system: carbohydrates – 172 kg/yr per person and major elements as urea and salts – 60–65 kg/yr per person.

No evidence confirming that BHB can be included in human diet has been obtained so far. This, however, is not necessary. Biomass should actually be fractionated and its components added to human diet (amino acids, proteins, major, and trace elements).

TABLE 3.1 Mutual Supply of Biogenic Elements and Food Components by the Human and the "Hydrogen Bacteria-Electrolysis" System

Component	Daily Requirement	Supply	Fulfillment of Requirement, %
System → The Human			
Oxygen, L	700	700	100
Water, L	3.9	3.9	100
Proteins, g	100–1200	470–500	500
Essential amino acids, g:	1–11	5.7	550
tryptophan	1–11	5.7	550
Leucine	6–12	43.0	600
Isoleucine	3–4	15.20	500
Threonine	2–3.6	22–25	1000
Valine	4	18.0	450
Lysine	3–5	36.0	1200

TABLE 3.1 *(Continued)*

Component	Daily Requirement	Supply	Fulfillment of Requirement, %
Methionine	2–4	8.6	210
Phenylalanine	2–4	21.6	500
Carbohydrates, g	400–500	14–18	3.5
Fats, g	25–30	21–28	100
Vitamins, mg:			
B_1	2	0.43	21
B_2	2	34.50	1700
C	75–100	122.40	120
Folic acid	1–2	17–23	1000
Mineral elements, g:			
Phosphorus	1.0–1.5	16.6	1000
Potassium	2.5–5.0	2.60	50
Sodium	4.0–6.0	0.36	0.06
Calcium	0.8	0.57	72
Magnesium	0.3–0.5	4.00	800
Chlorine	6.0–9.0	0	0
Iron	$150 \cdot 10^{-4}$	1.440	9600
Zinc	$150 \cdot 10^{-4}$	0.045	300
Manganese	$100 \cdot 10^{-4}$	0.55	50
Copper	$20 \cdot 10^{-4}$	0.022	1000
Molybdenum	$5 \cdot 10^{-4}$	0.024	4800
Fluorides	$10 \cdot 10^{-4}$	0	0
Iodides	$2 \cdot 10^{-4}$	0	0
The Human → System			
Carbon dioxide, L	575	575	100
Water, L	5.9	5.9	100
Mineral elements, g:			
Nitrogen	72.00	14–15	20.0
Phosphorus	16.60	2.0–2.5	13.4
Magnesium	4.00	0.1–0.3	7.5
Potassium	2.60	1.5–2.0	60.0
Calcium	0.576	0.1–0.3	17.0
Sodium	0.36	8–12	2220

Thus, it was proven that hydrogen bacteria are tolerant to the spaceflight conditions and that the "hydrogen bacteria-electrolysis" system is capable of solving the main tasks of human life support by completely regenerating oxygen and water in the system, utilizing carbon dioxide and human liquid wastes, and producing part of food substances.

More than a century ago, the great dreamer K. E. Tsiolkovsky wrote that "... humanity will not always stay on the Earth ... In pursuit of light and space it will first timidly step over the boundary of the atmosphere so as to conquer the whole space around the Sun later." At the present time, the interest in long-duration space flights is growing, and this may also revive the interest in hydrogen bacteria.

3.2.2   BIOMASS OF HYDROGEN BACTERIA (BHB): POTENTIAL PROTEIN SOURCE

Earth's resources are limited. Based on their division into renewable and nonrenewable, some forecasts predict a new industrial crisis in the middle of the 21st century, due to energy and raw material shortages. The currently used agricultural technologies are extensive and there is no reason to expect a fundamental breakthrough in their development. Although grain production has grown considerably due to advances in genetics, selection, and physiology of plants (the first wave of the "Green Revolution"), the increase in food supply still remains an issue of importance. One billion people living in industrially developed countries consume too many calories-3350 cal/d, while the rest of the world population consumes too few calories-about 2520 cal/d on the average. The largest part of the population in Africa and Asia consume 2040 cal/d on the average, which is below the physiological norm. Presumably, by the end of the 21st century, the world population may reach 10–12 billion. Overall mechanization is not a way to significantly increase grain production. Intensification of land use enhances adverse effects-soil erosion. Thus, one of the main tasks of scientific and technological progress is to find effective ways to increase resources of protein substances.

An unconventional and fundamentally new way to produce protein is microbiological synthesis, which uses material resources and energy more efficiently than conventional agricultural technologies, does not require large land areas, and is not affected by climate or weather conditions. Microbiological technologies are environmentally friendly as they do not use pesticides. The growth rate of microorganisms is several hundred and several thousand times higher than those of agricultural crops and animals, respectively. Microbial biomasses are qualitatively similar to animal proteins. Products of microbiological synthesis used to prepare animal feed improve the quality and assimilability of traditional vegetable feed; moreover, 1 t of nutrient yeast can save 5 t of grain and increase the efficiency of livestock production by 15–30%. An average modern factory manufacturing microbial protein, which produces 50 t/yr and occupies the area 0.2 ha, can satisfy protein requirements of 10 million people. To attain the same goal, agricultural technologies need either 16 000 ha of cropland or a farm producing 400 pigs daily. Thus, it is only natural that the microbiological way of producing protein substances has become one of the very promising technologies. In the late 1960s-early 1970s, research in this area was pursued worldwide. The term "single-cell protein" (SCP) emerged at that time, referring to whole dead dried microbial cells (algae, yeasts, bacteria, fungi) to be used as protein sources in human foods or animal feeds. The term is not quite exact as, in addition to proteins, microbial biomasses contain considerable amounts of other components-sugars, lipids, and nucleic acids.

Microbiological production is based on using biochemical activity of microorganisms that can assimilate a wide range of chemical compounds and synthesize biomass of high biological value. The technological process of microbiological production generally includes the following steps: production and preparation of feedstock, preparation of inoculum, fermentation, extraction, and concentration of microbial biomass, and drying of the product. The most significant technical and economic parameters of microbiological synthesis of protein are the cost per unit and the cost of the feedstock (50% and more of all production costs) and energy costs (15–30%).

Chemolitho-autotrophic microorganisms, hydrogen-oxidizing bacteria in particular, are promising protein producers. Hydrogen bacteria are interesting because they are autotrophic. If they are used as protein producers, biomass production will not depend on sources of organic material. This can be an opportunity for effective transformation of electrical and, in the future, atomic, and even solar, energy into protein through electrolytic, thermal, or photochemical decomposition of water. Calculations show that with the available free water resources of the Global Ocean, electrolysis followed by microbiological synthesis from hydrogen can provide enough protein for 100 billion people, which is the predicted global population in 300–400 years. Development of hydrogen engineering will make this type of microbiological synthesis, unrelated to the quantity of organic resources and releasing no harmful by-products or wastes to the environment, a very promising approach.

SCP as a feed product must comply with a number of specific requirements. The main requirements are high nutritional value, digestibility, and economic efficiency. The nutritional value of microbial protein, determined by its chemical composition is close to that of traditional protein food. Microbial biomass is nutritious if its components can be broken down by digestive enzymes of higher animals or the human. An obstacle to this can be cell walls of some producers, which have to be destroyed first and high levels of nucleic acids. The latter are metabolized in animal organisms and are removed with urine; hence, they are not hazardous to higher animals. For the human body high levels of nucleic acids are harmful, because their assimilation can cause metabolic dysfunction and morbid conditions. Thus, microbial biomass intended for food must be pretreated, using various methods of destruction and removal of nucleic acids.

Studies performed at the Institute of Biophysics SB RAS were comprehensive and consistent, including detailed chemical investigations of the biomass of hydrogen bacterium *A. eutrophus*, preliminary *in vitro* estimation of its biological value, and physiological-biochemical, zootechnical, veterinary, and toxicological studies, using higher animals, both in laboratory and in agricultural facilities [46]. The comparison shows that proteins of hydrogen bacteria, yeast, and microalgae are similar to casein in their amino acid composition. However, while the content of essential amino acids of hydrogen bacteria is similar to that of yeast, the total amounts of proteins in the biomasses (% DW) differ significantly. For yeast, this value (Ntotal × 6.25) averages 50% and for hydrogen bacteria 70%. Hydrogen bacteria proteins are rich in essential amino acids (their total amount reaches 40%).

Results of comparing amino acid compositions of hydrogen bacteria proteins [17] with the literature data on proteins of unicellular algae, yeast, and a true animal protein (casein) are listed in Table 3.2. The total fraction of hydrogen bacteria contains less lysine but more valine, methionine, phenylalanine, and tryptophan than meat.

An important parameter determining biological value of the protein is its fractional composition. Results of comparing protein fractions [46] and their amino acid contents in hydrogen bacteria, meat, and wheat grain are presented in Table 3.3. Separation into fractions is based on different solubility of cell proteins in salt and alkaline solutions. While the major part of meat proteins are contained in fractions most readily available to digestive enzymes-I and

II (82.7%), more than half of the total hydrogen bacteria proteins are found in Fractions III and IV (mostly III). These are structural proteins, which are extracted by alkali and are less available to proteases, i.e., less readily digestible than Fraction I and II proteins. Similar to proteins, essential amino acids of hydrogen bacteria are mostly contained in Fraction III.

TABLE 3.2 Amino Acid Composition of Proteins of Various Origins

Amino Acid, % Dry Weight	Samples of Hydrogen Bacteria			Yeast	Algae	Casein
	1	2	3			
Lysine	7.02	8.61	9.20	7.02	5.98	7.33
Histidine	1.96	2.48	1.40	1.96	1.81	2.20
Arginine	7.30	8.00	7.50	7.30	7.74	3.19
Aspartic	10.08	9.57	9.10	10.08	9.49	7.11
Threonine	5.29	4.52	5.20	5.29	4.88	4.22
Serine	4.02	3.47	4.80	4.02	4.86	5.72
Glutamic	12.56	11.17	9.30	12.56	13.12	22.20
Proline	4.58	3.46	0.80	4.58	5.74	10.41
Glycine	6.05	5.47	10.20	6.05	6.34	1.88
Alanine	9.07	8.80	13.40	9.07	9.18	2.96
Cystine	0.56	—	0.30	0.56	1.37	0.42
Valine	6.38	7.13	7.50	6.38	5.41	5.72
Methionine	2.63	2.69	0.40	2.63	2.16	2.47
Isoleucine	4.47	4.58	4.50	4.47	3.55	4.10
Leucine	8.60	8.52	8.70	8.60	8.91	9.39
Tyrosine	3.62	3.26	2.40	3.62	3.13	4.75
Phenylalanine	4.42	3.96	3.60	4.42	4.41	4.62
Tryptophan	1.40	—	—	1.40	1.58	1.32

(—) denotes "not analyzed."

Samples: 1 – [18]; 2 – [19]; 3 – [20]; yeast – [21].

One of the most significant factors for biological value of proteins is the degree to which they are attacked by proteolytic enzymes. Protease attackability of the studied proteins is investigated by comparing them with proteins that are readily assimilated by organisms of humans and animals. The study based on this approach takes into account the effects of the structural organization of cell components and pretreatment processes on protein availability.

TABLE 3.3   Fractional Composition of Proteins of Various Origins

Protein Source	Total Amount, % Dry Weight	Distribution of Protein Among Fractions*, % of Total Protein			
		I	II	III	IV
Meat	76.1 ± 1.2	30.5 ± 0.7	52.2 ± 0.7	16.6 ± 0.8	6.1 ± 0.1
Wheat	15.2 ± 0.8	25.9 ± 0.9	16.1 ± 0.9	39.0 ± 1.4	19.0 ± 0.4
Hydrogen bacteria	69.4 ± 1.2	28.0 ± 0.8	14.8 ± 0.8	52.8 ± 1.4	4.4 ± 0.1
Yeast**					
1	44.0 ± 1.4	10.0 ± 0.2	31.0 ± 0.1	33.0 ± 3.9	26.0 ± 1.6
2	48.0 ± 0.3	18.0 ± 2.2	31.0 ± 3.4	34.0 ± 3.7	17.0 ± 1.2

* Extractive agents: I – 0.03M KCl; II – 0.6M KCl + 0.04M $NaHCO_3$ + Na_2CO_3; III – 0.1M NaOH; IV – 1M NaOH at 65°C.

** Yeast: 1 – flow culture on alkanes, 2 – on glucose.

Attackability of protein fractions of hydrogen bacteria by pepsin and trypsin was investigated by Barashkov and his co-authors [22, 23]. The degree of proteolysis was estimated from the increase in amine nitrogen in the broth and expressed as a percentage. The amount of amine nitrogen formed during acid hydrolysis of samples in sealed ampoules containing 6H HCl for 22 h at 110° was taken as 100%. It was found that salt soluble proteins were more readily than casein attacked by pepsin and their digestibility due to sequential action of pepsin and trypsin was close to casein digestibility. Structural proteins sequentially attacked by pepsin and trypsin were digested poorly, not exceeding 50% of casein attackability.

A necessary step in evaluation of a new product of microbiological synthesis is testing of its safety and biological value in experiments with animals. Biological value of BHB produced on a Pilot Production Instal-lation in the Institute of Biophysics SB RAS (Krasnoyarsk) was tested experimentally on different agricultural animals both in laboratory and in agricultural facilities. These studies were performed by a large team of biotechnologists, chemists, physiologists, and medical specialists of the Institute of Biophysics SB RAS in cooperation with specialists of other research institutes and agricultural facilities. The work was organized and headed by Yuli N. Okladnikov, MD, Head of the Biomedical Team [24, 25].

Scientific and methodological assistance to these studies was provided by the All-Union Institute of Biosynthesis of Protein Substances of the USSR Ministry of Microbiological Industry and, personally, by Professor Nina B. Gradova.

The development of large-scale experiments on testing the biological value of the BHB became possible as a result of the creation of a pilot industrial production at the Institute of Biophysics of the Siberian Branch of the Russian Academy of Sciences, on which about 10 tons of biomass for testing in production conditions were obtained (Figure 3.3).

FIGURE 3.3 Pilot production of biomass fermenter V = 3000 L, and BHB – biomass hydrogen bacteria.

Source: Photo Tatiana Volova.

In experiments, animals that were fed the diet containing BHB were compared with control animals that were fed standard diets by analyzing their zootechnical, physiological, biochemical, and morphological-histological parameters. In the course of the experiments, animals were examined to monitor their growth, development, productivity, and health. The possibility and consequences of adding BHB to diets of agricultural birds and animals as animal protein component (instead of fish and meat-bone flours, whole milk, and skim milk) were investigated by researchers of the Institute of Biophysics SB RAS together with specialists of the Krasnoyarsk Agrarian University, the Krasnoyarsk Agricultural Research Institute, and the Krasnoyarsk Department of Research and Development Institute for Animal Breeding. Experimental animals were broiler chickens, egg-laying hens, pigs, calves, and fur-bearing animals (arctic foxes and minks) (Figure 3.4).

Thus, it was proven experimentally that BHB could be successfully used in animal feed as a source of animal protein. An important result of the experiments described above is that BHB used in tested amounts for tested periods of time did not cause any evident toxic effects. The results obtained in the experiments suggest that BHB can be used as part of the ration given to agricultural animals and birds, to replace traditional animal-derived feed. The range of this replacement is estimated as 25%–50% of the total animal protein in the feed (Figure 3.3), depending on what animal it is and on its age. The amount of the biomass added to the feed is determined by the type of animal or bird and the raising aims (reproduction, rapid growth, or fattening).

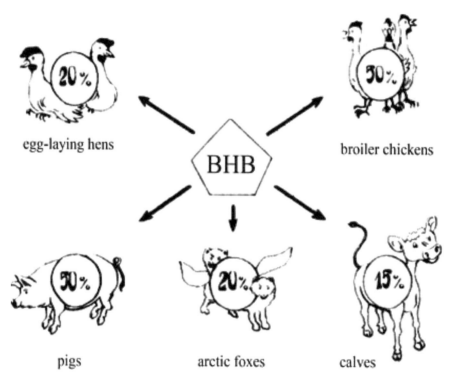

egg-laying hens

broiler chickens

BHB

pigs   arctic foxes   calves

FIGURE 3.4 Permissible percentages of biomass of hydrogen bacteria (BHB) used instead of animal proteins.

3.2.3 HYDROGEN BACTERIA AND SYNTHESIS OF BIODEGRADABLE POLYMERS

Growing production of synthetic plastics, including packaging materials, is a global environmental issue. Most of the packaging is now buried in landfills; not more than 10–15% is recycled in industrially developed countries. The projects of reusing of chemical plastics offer no grounds for optimism. Thus, a radical solution to the problem of "polymer garbage" is to create and use polymers that are, under appropriate conditions, biodegraded into components innocuous to living and non-living nature. Development and use of new, environmentally friendly materials, which will be able to be degraded in the environment without producing toxic compounds, thus joining the global material cycles, is among the priorities for critical technologies of the 21st century. Accumulation of synthetic plastics in the biosphere is a global ecological problem. The production of synthetic plastics has exceeded 350 million tons in year; 50% it is it is packaging. More than 99% of plastics are produced from chemicals derived from oil, natural gas, and coal, and all of which are dirty, non-renewable resources. Researchers estimate that more than 8.3 billion tonnes of chemical plastic has been produced since the early 1950s: only 10–15% of all plastic waste ever produced has been recycled; about 5% has been incinerated, while the rest > 80% has accumulated in landfills, dumps or the natural environment.

Now have seen a growing interest in studies on biopolymers (polymers of biological origin). There are two major kinds of biopolymers: polymers produced by biological systems (such as microorganisms) and chemically synthesized polymers based on biological feedstocks (amino acids, sugars, fats). In the past 15 to 20 years, biopolymer engineering has become one of the major interdisciplinary research areas. The main goal of this research is to: (i) find and study new biopolymers, and (ii) develop a fundamental basis for constructing biological systems capable of synthesizing polymers with target properties.

Among the biodegradable polymers that have already been developed or are being developed now for various applications, including medical ones, are aliphatic polyesters, polyamides, segmented polyester urethanes, polymers of lactic and glycolic acids (polylactides and polyglycolactides), silicon, polyethylene terephthalate, and, since recently, polymers of hydroxyalkanoic fatty acids, the so-called PHAs [42, 26–31].

PHAs are natural macromolecules (polymers of hydroxy-derived fatty acids) synthesized by prokaryotic organisms under specific conditions of unbalanced growth as intracellular storage of energy and carbon from a variety of substrates (sugars, organic acids, alcohols, etc.). The list of microorganisms capable of producing different yields of PHAs is growing rapidly, now containing more than 300 organisms. However, just a few species of microorganisms have been chosen for commercial synthesis. Among them are the chemoorganotrophic organism *j* (until recently known as *Ralstonia* and *Alcaligenes*), which can use different carbon sources, and heterotrophic bacteria of three taxa: *Methylotrophus*, *Methylobacterium*, and *Pseudomonas.*

The efficiency of P(3HB) production from hydrogen as an energy substrate is very high; the yield coefficient of polymer production from hydrogen amounts to 1.0. The use of the poorly soluble and explosive substrate is the major technological challenge for the implementation of this process.

Due to their specific growth physiology and constructive metabolism, in the middle of the linear growth phase, hydrogen-oxidizing bacteria stop synthesizing protein and begin accumulating PHAs even on complete nutrient medium. Based on this fact, we investigated the cell growth and polymer yields in autotrophic batch culture of *C. eutrophus* B-10646 with different inflowing amounts of nitrogen in a 10-L fermenter, with the working volume 3 L. [34]. In the experiment with the amount of the nitrogen supplied to the culture corresponding to the physiological requirement of the cells (120 mg/g cells), for 70 h of cultivation, the intracellular P(3HB) content had reached 55% and cell concentration 30 g/L (Figure 3.5a).

The second cultivation (Figure 3.5b) mode consisted of two stages: in the first, the cells were grown on complete nutrient medium with continuous inflow of nitrogen in the amounts corresponding to the physiological requirement of cells (120 mg/g cells); in the second the cells were grown on nitrogen-free medium. Under these conditions, the polymer yield reached 80%, but the duration of the process was increased to 80–85 h. In the third mode of batch cultivation, cells were grown under limited nitrogen supply in the first stage and without nitrogen at all in the second. The best results were obtained with nitrogen supply amounting to 50% of that required by cells in the first stage (60 mg/g cells) (Figure 3.5c). By the end of the experiment (70 h), intracellular polymer content had reached 85% and cell concentration 48 g/L.

In order to achieve biosynthesis of PHA copolymers, one needs to supplement the culture medium with additional carbon substrates that are toxic to bacteria (valerate, hexanoate, butyrolactone, etc.). Therefore, it is important to know maximum tolerable concentrations of precursor substrates for each specific culture and the mechanisms of formation of copolymers. As a substrate of endogenous respiration, PHA is metabolized intracellularly. Polymer carbon chains are broken down, and monomers containing more than 4 carbon atoms get involved in PHA resynthesis as shorter monomers, mainly 3-hydroxybutyrate. As different monomer units are synthesized and incorporated into PHAs with different rates, the duration of cultivation after the addition of the second precursor C-substrate should be regulated, or else the fraction of the second monomer in the PHA would be decreased [35].

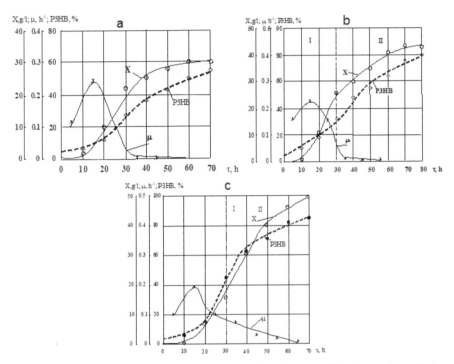

FIGURE 3.5 Parameters of *C. eutrophus* B-10646 batch autotrophic culture: (a) a single-stage process on complete medium; (b) a two-stage process: Stage I (I) on complete medium and Stage II (II) on nitrogen-free medium; (c) a two-stage process: Stage I (I) at 50% nitrogen supply and Stage II (II) on nitrogen-free medium; X-cell concentration of the culture (g/L), μ-cell specific growth rate, h^{-1}; P(3HB)-intracellular polymer concentration (% of dry matter).

Enhanced tolerance of *Cupriavidus eutrophus* B-10646 to valerate, hexanoate, and γ-butyrolactone compared with *R. eutrophus* B 5786 enabled its cultivations under conditions leading to the production of copolymers with different chemical structure. Autotrophic cell culture was supplemented with precursor C-substrates such as 3-hydroxyvalerate (3HV), 3-hydroxyhexanoate (3HHx), or 4-hydroxybutyrate (4HB) in amounts not exceeding the maximum tolerable concentrations (3–5 g/L) for the study strain, and, as a result, PHAs with different chemical structures were synthesized (Table 3.4).

Cultivation of *C. eutrophus* B 10646 on mixed carbon substrate containing CO_2 and potassium valerate yielded copolymers with major fractions of 3-hydroxybutyric and 3-hydroxyvaleric acids. By controlling the inflow of valerate and the duration of the cultivation following the addition of this substrate, polymers with different molar fractions of 3HB and 3HV were synthesized: from 36 to 93.5% and from 6.5 to 85%, respectively. Synthesis of copolymers containing 3HHx is a more complex task as this precursor substrate is more toxic for bacteria. The addition of potassium hexanoate resulted in the formation of PHAs that mainly contained 3-hydroxybutyrate (79.3 to 94.1 mol%), with the 3HHx fraction making up 5 to 20 mol% and a minor fraction of 3HV (0.7–1.05 mol%). A fourth PHA synthesized in these experiments was a copolymer containing major molar fractions of 3-hydroxybutyric and 4-hydroxybutyric acids. To synthesize these copolymers, the culture medium was supplemented with γ-butyrolactone (or 1,4-butanediol). The resulting polymers contained different molar fractions of 3HB and 4HB (from 10.0 to 51.3%) and minor molar fractions of 3HV (0.29–0.48%) and 3HHx (0.04–0.41%) (Table 3.4).

The tolerance of *C. eutrophus* B10646 to the salts of alkanoic acids and γ-butyrolactone made it possible to implement a growth cycle of this strain under autotrophic conditions with the culture medium supplemented with two precursor substrates: potassium propionate and γ-butyrolactone or potassium valerate and potassium hexanoate. As a result, for the first time, *C. eutrophus* B10646 cells cultured autotrophically synthesized PHAs with different compositions, which contained major molar fractions of 3-hydroxybutyrate (25.3 to 84.89%), 4HB (23.6 to 62.5%), 3HV (1.25 to 24.9%), and 3HHx (2 to 25.1%) (Figures 3.6, Table 3.5). Some PHAs, synthesized in the culture supplemented with hexanoate, contained minor molar fractions of 3-hydroxyoctanoate (3HO).

Figure 3.6 shows ion chromatograms and mass spectra of monomer units constituting the polymers: PHA copolymer synthesized by bacterium *C. eutrophus B-10646* from mixed carbon substrate.

Results of comparative investigations of polymer properties using X-ray structure analysis, differential scanning calorimetry, and gel permeation chromatography are given in Tables 3.4 and 3.5. No clear relationship has been found between molecular mass of the polymers and their chemical structure, i.e., the molar fractions of their monomer units. The polymers containing major fractions of 3HB, 3HV, and 4HB, however, generally showed lower values of weight average and number average molecular weight (439–756 kDa and 134–183 kDa). At the same time, PHAs that contained major fractions of only two monomers (3HB and 4HB or 3HB and 3HV, or 3HB and 3HHx) showed Mw and Mn values reaching 837 and 209; 1336 and 500; 924 and 225 kDa.

TABLE 3.4 Characterization of PHA Copolymers Synthesized by Bacterium *C. Eutrophus* B-10646 from CO_2 Supplemented with One Precursor Substrate (Valerate or Hexanoate, or γ-Butyrolactone)

Carbon Source	PHA Composition, mol%				T_g	T_m	T_d	C_x	Mw, kDa	Đ
	3HB	**4HB**	**3HV**	**3HHx**						
CO_2	99.10	0	0.90	0	—	179.0	294.8	76	922	2.51
CO_2 +	78.87	20.30	0.42	0.41	40.1	167.8	288.0	38	605	4.19
γ-butyro–	62.35	37.10	0.29	0.26	36.4	170.5	287.0	36	537	3.44
lactone	48.20	51.30	0.34	0.16	38.1	169.4	290.7	38	837	4.00
CO_2 +	80.00	0	20.00	0	56.8	175.1	282.9	57	1132	2.99
valerate	63.00	0	37.00	0	51.7	179.6	284.4	51	1336	2.67
	36.00	0	64.00	0	—	153.0	249.0	53	1111	3.49
	15.00	0	85.00	0	ND	ND	ND	50	1120	3.11
CO_2 +	83.95	0	1.05	15.00	—	174.8	283.8	60	924	4.10
hexanoate	79.28	0	0.72	20.00	—	164.0	256.0	42	421	3.02

ND: not determined; dash (—): T_g is not detected.

The decrease in the molecular weight of a PHA caused a decrease in its glass transition temperature, melting point, and thermal degradation temperature. The incorporation of 4HB, 3HV or 3HHx into the poly(3-hydroxybutyrate) chain influenced the temperature parameters of the polymers, too; the general trend was for the temperature parameters to decrease with an increase in the molar fractions of these monomer units. The difference between the melting point and the thermal degradation temperature stayed within a range of 100–120°C, i.e., the polymers remained thermally stable.

The degree of crystallinity of polymers (C_x) was definitely found to be related to the composition and molar fractions of monomer units of PHAs. The degree of crystallinity of PHA copolymers is significantly lower than that of homogenous P(3HB) (75 ± 5%). The influence of the monomer units on the ratio of crystalline to amorphous phase in copolymers containing major fractions of two monomer units increased from 3HV to 3HHx to 4HB. The properties of PHAs containing major fractions of three monomer units

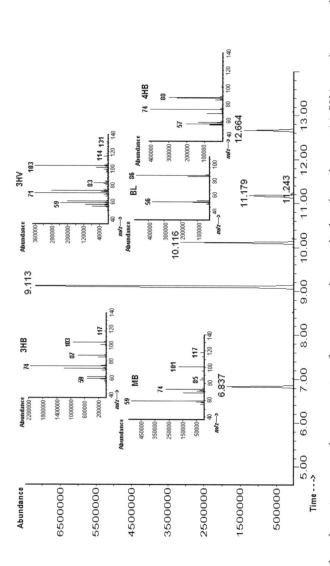

FIGURE 3.6 Ion chromatograms and mass spectra of monomer units constituting the polymers: (a) PHA copolymer synthesized by bacterium *C. eutrophus* B-10646 from mixed carbon substrate CO_2 + γ-butyrolactone + propionate: -methyl-4-metoxybutyrate (MB)-6.837; methyl-3-hydroxybutyrate (3HB)-9.113; methyl-3-hydroxyvalerate (3HV)-10.116; butyrolactone (BL)-11.179; methyl-3-hydroxyhexanoate (3HHx)-11.243; methyl-4-hydroxybutyrate (4HB)-12.664 min; (b) PHA copolymer synthesized by bacterium *C. eutrophus* B-10646 from mixed carbon source CO_2 + valerate + hexanoate: -methyl-3-hydroxybutyrate (3HB) with retention time 9.127; methyl-3-hydroxyvalerate (3HV)-10.108; methyl-3-hydroxyhexanoate (3HHx)-11.251; methyl-3-hydroxyoctanoate (3HO)-13.705 min.

(3HB/4HB/3HV or 3HB/3HV/3HHx) are generally similar to those of two-component PHAs; as the fractions of 4HB, 3HV and/or 3HHx are increased, their temperature characteristics (glass transition temperature, melting point, and thermal degradation temperature) become 10–15°C lower than those of P(3HB). The degree of crystallinity decreases to 35–40% in 3HB/3HV/3HHx and to 8–22% in 3HB/4HB/3HV.

TABLE 3.5 Characterization of PHA Copolymers Synthesized by Bacterium *C. Eutrophus* B-10646 from CO_2 Supplemented with Two Precursor Substrates (γ-Butyrolactone + Propionate or Hexanoate + Valerate)

Carbon Source	PHA Composition, mol%				T_g	T_m	T_d	Cx	Mw, kDa	Đ
	3HB	4HB	3HV	3HHx						
CO_2	99.10	0	0.90	0	—	179.0	294.8	76	922	2.5
CO_2 + propionate + γ-butyro-lactone	59.45	34.50	5.40	0.65	45.9	161.3	295.3	21	439	3.0
	25.30	62.50	12.20	0	44.0	163.5	295.2	18	508	3.4
	56.20	23.60	20.20	0	44.0	171.4	280.5	22	645	3.7
	48.30	34.80	16.90	0	36.0	171.8	283.8	8	475	3.5
	55.80	31.30	13.30	0.12	43.0	167.8	283.3	21	756	4.1
CO_2 + valerate + hexanoate	73.10	0	24.90	2.00	47.8	173.4	268.8	54	537	3.6
	75.90	0	13.00	11.10	ND	166.0	265.0	35	999	3.0
	84.89	0	1.25	12.70	58.0	167.8	221.0	65	288	2.2
	67.30	0	7.20	25.10	ND	164.0	263.0	40	970	3.5

ND – not determined.

Thus, by cultivating *C. eutrophus* B10646 in the two-stage batch mode, without essentially changing the process, with varied conditions of carbon nutrition of the cells grown using the $CO_2/O_2/H_2$ gas mixture as the main growth substrate, we employed the methods for production of PHAs with different molar fractions of monomer units and synthesized the polymers that had significantly different physicochemical properties.

On the basis of experimentally achieved and developed technical and technological parameters of biosynthesis processes, the baseline data were obtained and the project was worked out for organization of pilot production of degradable PHAs. The pilot line for production of PHAs was designed, equipped, and commissioned, including the "bioengineering" fermentation line (Switzerland), as well as the blocks for preparation of inoculum and culture media, isolation, and purification of polymers. The pilot production was commissioned. Scaling the technology made it possible to obtain the experimental batches of polymers in the amounts needed for conducting a complex of prescribed research, development, and standardization of specialized polymer products [36].

PHA successfully used to produce various types of polymer products (Figure 3.7). As a result of the development of synthesis processes for new types of PHA and using them, experimental products were designed and studied in the form of films, barrier membranes, granules, filling material, ultra-thin fibers obtained by electrostatic molding, bulk dense and porous implants for reconstruction of bone tissue defects, in the form of tubular endobiliary stents, mesh endoprostheses modified with a PHA coating, microparticles for the deposition and delivery of drugs, etc. [47].

There are two possible ways to broaden PHA applications. One way is to develop large-scale PHA production, i.e., to increase PHA outputs and reduce their cost by manufacturing inexpensive items such as packaging materials, everyday articles, films, and pots for agriculture, etc. The other way is to establish small-scale facilities for the production of high-cost specialized items. PHAs show the greatest potential in medicine and pharmacology.

In Russia, biomedical studies of PHAs and experimental prototypes of PHA products were initiated by the researchers of the Institute of Biophysics SB RAS and the Siberian Federal University. In cooperation with the Institute of Transplant Surgery and Artificial Organs (currently the V. I. Shumakov Federal Research of Transplant Surgery and Artificial Organs Center), the research team studied films of poly-3-hydroxybutyrate and P(3HB-co-3HV) copolymers and proved that they did not produce any cytotoxic effect when contacting directly with the cultured cells and when implanted as sutures in vivo; high-purity specimens were suitable for contact with blood. Results of these studies were summarized in the first Russian book on PHAs: "PHAs-biodegradable polymers for medicine" [37]. Since then, the research team has considerably widened the scope of its PHA studies by synthesizing PHAs with different chemical compositions and by designing and studying experimental films, barrier membranes, granules, filling materials, ultrafine fibers produced by electrospinning, solid, and porous 3D implants for bone tissue defect repair, tubular biliary stents, mesh implants modified by PHA coating, microparticles for drug delivery, etc. In cooperation with the V. F. Voino-Yasenetsky Krasnoyarsk Federal Medical University, the research team has conducted pioneering clinical trials. Results have been covered by several Russian patents, reported in papers published in peer-reviewed Russian and international journals, summarized in books and reviews [33, 38–40].

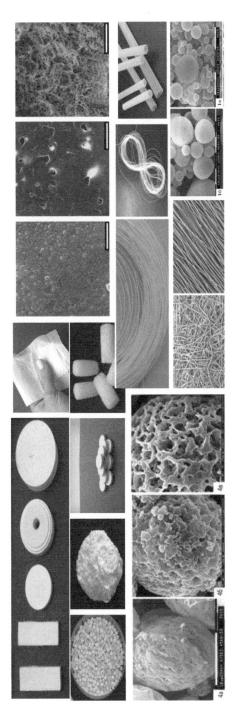

FIGURE 3.7 PHA products.

Source: Photo T. Volova.

An innovative direction in the use of destructible bioplastics has been implemented – the design of prolonged and targeted products for the protection of cultivated plants. Based on our knowledge of polyhydroxyalkanoate synthesis and kinetics of degradation in natural ecosystems and techniques developed for processing polymers into specialized products, we have formulated an innovative direction in using PHA. The effectiveness of fungicidal and herbicidal preparations, differing in biodegradation rates and the kinetics of the release of active substances, has been proven in crops of cultivated plants infected with fusarium and weeds. The developed forms provide targeted and controlled delivery of drugs to plants; their use is intended to reduce the rates of application and the risk of uncontrolled spread of xenobiotics in the biosphere. The obtained pioneering results are generalized in a monograph "New Generation Formulations of Agrochemicals: Current Trends and Future Priorities" [48].

The brief review of the results presented in the chapter indicates the high biotechnological potential of hydrogen-oxidizing bacteria, which are rightfully promising and promising objects of biotechnology.

ACKNOWLEDGMENTS

This study was financially supported by programs and research plans of the Siberian Branch of the Russian Academy of Sciences; series of grants of the Russian Foundation for Basic Research, as well as Projects: "Biotechnologies of Novel Biomaterials"(Agreement No.: 11.G34.31.0013) 2010–2014 and "Agropreparations of the new generation: a strategy of construction and realization" (No 074-02-2018-328) 2018–2019 in accordance with Resolution No 220 of the Government of the Russian Federation of April 9, 2010, "On measures designed to attract leading scientists to the Russian institutions of higher learning."

KEYWORDS

- **assimilation quotient**
- **biomass of hydrogen bacteria**
- **life support systems**
- **polyhydroxyalkanoates**
- **respiratory quotient**
- **single-cell protein**

REFERENCES

1. Kaserer, H., (1906). The oxidation of the water material by microorganisms. *Zbl. Bacteriol. II. 16*(5), p. 681.

2. Lebedeff, A. F., (1907). *About the assimilation of the carbon by hydrogen oxidizing bacteria. Biochim., 7,* 1–16.

3. Nabokish, A., & Lebedeff, A. F., (1906). Fiber through the oxidation of the hydrogen bacteria. *Zbl. Bacteriol. Parasites. Infect. Illness and Hyd., Abbot, 17,* 350.

4. Zavarzin, G. A., (1972). *Litotrophik Mikroorganizmi (Litothrophic Microorganisms).* Nauka, Moscow (in Russian).

5. Bongers, L. H., (1964). Sustaining life in space: A new approach. *Aerospace Med., 139*(1), 144–162.

6. Jenkins, D. W., (1965). Electrolysis hydrogenomonas bacterial bioregenerative life support system. *Proc. XVI Intern. Astronaut. Congress* (pp. 229–244). New York.

7. Voronin, G. I., & Polivoda, A. I., (1967). *Zhizneobespecheniye Ekipazhei Kocmicheskikh Korablei (Life Support of Spacecraft Crews)* (p. 210). Moscow: Mashinostroyeniye. (In Russian).

8. Terskov, I. A., Gitelson, I. I., Sidko, F. Y., et al., (1973). Culture of hydrogen-oxidizing bacteria as a promising producer of protein for terrestrial uses and manmade ecological life support systems. In: *Abstracts of the 24th International Astronautic Congress* (pp. 395–402). Moscow. (In Russian).

9. Terskov, I. A., Gitelson, I. I., Sidko, F. Y., Trubachev, I. N., Fyodorova, Y. V., Volova, T. G., & Batutin, M. E., (1977). Physiological and biochemical characterization of continuous culture of hydrogen bacteria. In: *Fundamentalnyye Issledovaniya (Fundamental Studies)* (pp. 32–37). Novosibirsk: Nauka. (in Russian).

10. Terskov, I. A., Sidko, F. Y., Gitelson, I. I., Okladnikov, Y. N., Volova, T. G., & Fyodorova, Y. V., (1984). On biological and nutritional value of biomass of hydrogen bacteria. *Prikladnaya Biokhimiya i Mikrobiologiya (Applied Biochemistry and Microbiology), 20*(4), 510–520. (In Russian).

11. Volova, T. G., Fyodorova, Y. V., & Trubachev, I. N., (1980). Hydrogen bacteria and life support systems. In: *Proizvodstvo Belka na Vodorode (Protein Production from Hydrogen)* (pp. 113–123). Novosibirsk: Nauka. (In Russian).

12. Volova, T., Gitelson, J., Terskov, I., & Sidko, F., (1999). Hydrogen bacteria as a potential regenerative LSS component and producer of ecologically clean degradable plastic. *Life Support and Biosphere Science, 6,* 209–213.

13. Volova, T. G., (2009). *Hydrogen-Based Biosynthesis* (p. 287). Nova Science Pub. Inc. NY, USA.

14. Volova, T. G., Trubachev, I. N., & Veber, M. I., (1975). Continuous cultivation of hydrogen bacteria with repeated use of the medium.*Mikrobiologiya (Microbiology), 44*(2), 219–225 (In Russian).

15. Trubachev, I. N., Kalacheva, G. S., Andreeva, R. I., & Voitovich, Y. V., (1971). Biochemical composition of hydrogen bacteria depending on growth conditions. *Mikrobiologiya (Microbiology), 40*(3), 424–427. (In Russian).

16. Volova, T. G., (1974). Mineralnoye pitaniye u vodorodnykh bakterii (Mineral nutrition of hydrogen bacteria). *Summary of PhD Thesis* (p. 24). IM.-Moscow. (in Russian).

17. Andreeva, R. I., & Volova, T. G., (1973). Amino acid composition of hydrogen bacteria depending on the conditions of mineral nutrition. *Prikladnaya biokhimiya i mikrobiologiya (Applied Biochemistry and Microbiology), 9*, 549–553 (in Russian).

18. Trubachev, I. N., & Andreeva, R. I., (1969). Amino acid composition of hydrogen bacteria in continuous culture. *Informatsionnyi Byulleten (Information Bulletin),5*, 119–120. Irkutsk (in Russian).

19. Calloway, D. H., & Kumar, E. M., (1969). Protein quality of the bacteria *Hydrogenomonas eutropha. Appl. Microbiol., 17*, 176–178.

20. Vedenina, I. Y., (1968). Avtotrofnaya assimilyatsiya uglekisloty vodorodnymi bakteriyami *Hydrogenomonas* ZI (Autotrophic assimilation of carbon dioxide by the autotrophic bacterium *Hydrogenomonas* ZI). *PhD Thesis (Biology)*, (p. 130). IM.-Moscow. (In Russian).

21. Pokrovskii, A. A., & Somin, V. I., (1972). Amino acid composition of some unicellular organisms. In: *Mediko-Biologicheskiye Issledovaniya Uglevodorodnykh Drozhzhei (Biomedical Studies of Hydrocarbon Yeasts)* (pp. 103–109). Moscow (in Russian).

22. Barashkov, V. A., Trubachev, I. N., & Gitelson, I. I., (1976). Comparative characterization of amino acid compositions of protein fractions of hydrogen bacteria. *Voprosy pitaniya (Dietary Issues),1*, 73–75. (in Russian).

23. Barashkov, V. A., Trubachev, I. N., & Gitelson, I. I., (1977). Proteolysis of proteins of hydrogen bacteria and *Chlorella* by pepsin and trypsin. *Mikrobiologicheskii Zhurnal (The Journal of Microbiology), 39*, 498–499. Kiev (in Russian).

24. Okladnikov, Y. N., (1980). Biomass of hydrogen bacteria used as part of the diet of farm animals and birds. In: *Proizvodstvo Belka na Vodorode (Protein Production from Hydrogen)* (pp. 94–112). Novosibirsk: Nauka. (in Russian).

25. Gitelson, I. I., Volova, T. G., Okladnikov, Y. N., Sidko, F. Y., Terskov, I. A., Trubachev, I. N., & Fyodorova, Y. V., (1981). *Biomassa Vodorodnykh Bakterii-Novyi Istochnik Kormovogo Belka (Biomass of Hydrogen Bacteria: A New Source of Feed Protein)* (p. 50). Preprint of IPSB-20B.-Krasnoyarsk. (in Russian).

26. Sudesh, K., Abe, H., & Doi, Y., (2000). Synthesis, structure, and properties of polyhydroxyalkanoates: Biological polyesters. *Prog. Polym. Sci., 25*, 1503–1555.

27. Chen, G. Q., (2009). A microbial polyhydroxyalkanoates (PHA) based bio-and materials industry. *Chem. Soc. Rev., 38*, 2434–2446.

28. Chen, G. Q., (2010). Industrial production of PHA. Microbiol. Monogr. In: Chen, G. Q., & Steinbüchel, A., (eds.), *Plastics from Bacteria: Natural Functions and Applications* (pp. 121–132). Springer.

29. Chen, G. Q., & Jiang, X. R., (2017). Engineering bacteria for enhanced polyhydroxyalkanoates (PHA) biosynthesis. *Synth. Syst. Biotechnol., 2*, 192–197.

30. Koller, M., Maršálek, L., Dias, M. S. M., & Braunegg, G., (2017). Producing microbial polyhydroxyalkanoate (PHA) biopolyesters in a sustainable manner. *New Biotechnol., 37*, 24–38.

31. Koller, M., (2018). Chemical and biochemical engineering approaches in manufacturing polyhydroxyalkanoate (PHA) biopolyesters of tailored structure with focus on the diversity of building blocks. *Chem. Biochem. Eng. Q., 32*, 413–438.

32. Volova, T. G., (2004). *Polyhydroxyalkanoates-Plastic Material of 21st Century (Production, Properties, Application)* (p. 292). Nova Science Pub. Inc., NY.

33. Volova, T., Kiselev, E., Shishatskaya, E., Zhila, N., Boyandin, A., Syrvacheva, D., Vinogradova, O., et al., (2013). Cell growth and PHA accumulation from CO_2 and H_2 of

a hydrogen-oxidizing bacterium, *Cupriavidus eutrophus* B-10646. *Bioresour. Technol.*, *146*, 215–222.

34. Volova, T. G., Shishatskaya, E. I., & Sinskey, A. J., (2013). *Degradable Polymers: Production, Properties, Applications*. Nova Science Pub. Inc., NY.

35. Volova, T. G., Kalacheva, G. S., & Steinbüchel, A., (2008). Biosynthesis of multi-component polyhydroxyalkanoates by the bacterium *Wautersia eutropha. Macromol. Symp., 269*, 1–7.

36. Kiselev, E. G., Demidenko, A. V., Baranovskiy, S. V., & Volova, T. G., (2014). Scaling of biodegradable polyhydroxyalkanoates synthesis technology in pilot production conditions. *Journal of Siberian Federal University, Biology, 2*, 134–147.

37. Volova, T. G., Sevastianov, V. I., & Shishatskaya, E. I., (2006). *Polyhydroxyalkanoates-Biodegradable Polymers for Medicine* (p. 237). Krasnoyasrs-Platina.

38. Volova, T., Goncharov, D., Nikolaeva, E., & Shishatskaya, E., (2017a). Chapter: Electro spinning of degradable PHAs: Process, properties, applications. Nova Science Publishers, Inc. (N.Y.) In: *Book: Electro Spinning: Fundamentals, Methods and Applications*(50c). Nova Science Publishers, Inc. (N.Y.) USA.

39. Volova, T. G., Vinnik, Y. S., Shishatskaya, E. I., Markelova, N. M., & Zaikov, G. E., (2017b). *Natural-Based Polymers for Biomedical Applications* (p. 460). Canada: Apple Academic Press.

40. 40. Volova, T. G., & Shishatskaya, E. I., (2014). Results of biomedical studies of PHAs produced in the Institute of Biophysics SB RAS and Siberian Federal University. *Chapter in the Book: Polyhydroxyalkanoates (PHA): Biosynthesis, Industrial Production and Applications in Medicine*. Nova Sciences Publ. Inc. NY. USA.

41. Gitelson, I. I., (1980). *Proizvodstvo Belka na Vodorode (Protein Production from Hydrogen)* (p. 150). Novosibirsk: Nauka. (in Russian).

42. Schlegel, H. G., Gottschalk, G., & Von, B. R., (1961). Formation and utilization of poly-β-hydroxybutyric acid by Knallgas bacteria (*Hydrogenomonas*). *Nature, 191*, 463–465.

43. Tkachenko, V. V., Rylkin, S. S., Shkidchenko, A. N., & Sterkin, V. E., (1971). The effect of culture conditions on the fractional composition of proteins of microorganisms. *Mikrobiologiya (Microbiology), 40*, 651–655. (In Russian).

44. Volova, T. G., Shishatskaya, E. I., Prudnikova, S. V., Zhila, N. O., & Boyandin, A. N., (2020). *New Generation Formulations of Agrochemicals: Current Trends and Future Priorities* (p. 286). Toronto–Canada: CRC/Taylor & Francis: Apple Academic Press.

45. Volova, T. G., Terskov, I. A., & Sidko, F. Y., (1985). *Mikrobiologicheskii Sintez na Vodorode (Microbiological Synthesis from Hydrogen)* (p. 148). Novosibirsk: Nauka. (In Russian).

46. Gitelson, J. (1980). *Proizvodstvo Belka na Vodorode (Protein Production from Hydrogen)* (In Russian) (149 p.). Novosibirsk: Nauka.

47. Volova, T. G., Vinnik, Yu.S., Shishatskaya, E. I., Markelova, N. M., & Zaikov, G. E. (2017). *Natural-Based Polymers for Biomedical Applications*. Canada: Apple Academic Press (460 p.)

48. Volova, T. G., Shishatskaya, E. I., Prudnikova, S. V., Zhila, N. O., & Boyandin, A. N. (2020). New Generation Formulations of Agrochemicals: Current Trends and Future Priorities. Toronto–Canada: CRC/Taylor & Francis: Apple Academic Press (p. 286) ISBN: 9781771887496. DOI 10.1201/9780429433610.

CHAPTER 4

Abutilon indicum, Prosopis juliflora, and *Acacia arabica* as Antibacterial Agents Against *Xanthomonas axonopodis* pv. punicae

AISHWARYA A. ANDHARE and RAVINDRA S. SHINDE

Department of Microbiology, Biotechnology, and Chemistry,
Dayanand Science College, Latur – 413512, Maharashtra, India,
E-mail: rss.333@rediffmail.com (R. S. Shinde)

4.1 INTRODUCTION

Pomegranate (*Punica granatum* L.) is an ancient fruit, belonging to the smallest botanical family punicaceae. Pomegranate is a native of Iran, where it was first cultivated in about 2000 BC, but spread to the Mediterranean countries at an early date. It is extensively cultivated in Spain, Morocco, and other countries around the Mediterranean, Egypt, Iran, Afghanistan, Arabia, and Baluchistan. Pomegranate is a good source of carbohydrates and minerals such as calcium, iron, and sulfur. It is rich in vitamin C and citric acid is the most predominant organic acid in pomegranate [1]. Apart from the fleshy portion of the fruit, the crop residues are also finding a place in industries. The rind of the fruit is a good source of dye, which gives yellowish-brown to khaki shades and is being used for dying wool and silk. The flower and buds yield light red dye, which is used for dying of cloths in India. The bark of the stem and root contains a number of alkaloids belonging to pyridine group. The bark is used as a tanning material especially in Mediterranean countries in the East [2].

Pomegranate is regarded as the "Fruit of Paradise." It is one of the most adaptable subtropical minor fruit crops and its cultivation is increasing very rapidly. In India, it is regarded as a "vital cash crop," grown in an area of

1,16,000 ha with a production of 89,000 MT with an average productivity of 7.3 MT [3]. Successful cultivation of pomegranate in recent years has met with different traumas such as pest and diseases. Among diseases bacterial blight caused by *Xanthomonas axonopodis pv. Punicae* is a major threat. Since 2002, the disease has reached the alarming stage and hampering the Indian economy vis-à-vis export of quality fruits. The disease accounted up to 70–100% during 2006 in Karnataka and Maharashtra resulting in wipeout of pomegranate. During the year 2007, the total output of pomegranate production in India was down by 60% [4].

The causal organism of blight is *Xanthomonas axonopodis* pv. *Punicae* [7] Vauterin, Hoste, Kersters, and Swings [5, 6]. The plant is susceptible to blight during all stages of growth and results in huge economic loss. Bacterial blight primarily affects the above-ground plant parts, especially leaves, twigs, and fruits. While the leaves how early water-soaked lesions to late necrotic blighting, the fruits show isolated or coalesced water-soaked lesions followed by necrosis with small cracks and splitting of the entire fruit [8]. Stems show lesions around nodes or injuries, forming cankers in later stages. Suspected symptoms on floral parts have also been reported [9, 10].

It is also presumed that the stem canker could be an outcomes of systemic spread of bacterium from leaf, although reported by Chand and Kishun [9] and Rani et al. [10] attempts to reproduce the field symptoms of blight on detached leaves, twigs, and fruits were unsuccessful in the artificial inoculations. The management of bacterial blight of pomegranate is a major concern. This disease could not be effectively managed with conventional antibiotics like streptomycin in field conditions. Continual and indiscriminate use of synthetic antibiotics to control bacterial disease of crop plants has caused health hazards in animals and humans due to their residual toxicity [11]. A bioactive principle isolated from plant appears to be one of an alternative for control of plant and human pathogens developed resistant to antibiotics. Plant originated-antibacterial compounds can be one approach to plant disease management because of their eco-friendly nature [12].

The ability of microorganisms to acquire and transmit resistance against antibiotics causes nosocomial and community-acquired infections [13]. The development of resistance in microorganisms to presently available antibiotics has necessitated the search for new antimicrobial agents [14]. In developing countries, about 80% of the population utilizes medicinal plants for the treatment of infectious diseases [15]. Indian gum Arabic tree *Acacia*, belong to the family leguminosae, and has been recognized worldwide as a multipurpose tree. *Acacia arabica* bark has been used as demulcent, nutritive supplement, Phytochemical screening of the stem bark of *Acacia arabica* revealed that the plant contains amines and alkaloids (dimethyltryptamine, 5-methoxy-dimethyltryptamine, and N-methyltryptamine), cyanogenic glycosides, cyclitols, saponins, fatty acids, and seed oils, fluoroacetate, gums, nonprotein amino acids, terpenes, hydrolyzable tannins, flavonoids, and condensed tannins [16]. Flavonoids, sterols/triterpenoids, alkaloids, and phenolics are known to be bioactive antidiabetic principles [17].

Prosopis juliflora (Sw.) DC (Fabaceae) is an evergreen tree native to South America, Central America, and the Caribbean. *Prosopis* species are generally fast-growing, drought-resistant, nitrogen-fixing trees or shrubs adapted to poor and saline soils in arid and semi-arid zones [18]. *P. juliflora* was first introduced to Kenya in 1973 for the rehabilitation of quarries and to safeguard the existing natural vegetation from overexploitation, but over the years, *Prosopis* has spread outside the designated plantation areas, adversely affecting natural habitats, rangelands, and cultivated areas in many parts of the country [19]. Because this is an exotic plant species that was introduced into the country, its use as a phytomedicine is not widespread and therefore the aim of this study was to determine the antibacterial activity of leaves and roots ethanolic extracts of *P. juliflora.* The syrup, made from ground pods of *P. juliflora*, is used for nourishment of underweight kids or those suffering from retardation in motor development. The syrup may raise lactation as well [20]. *Prosopis juliflora* (family Leguminosae, subfamily Mimosoideae) is probably the most widespread species, being a major source of fuel and fodder [21, 22].

Abutilon indicum belongs to the family Malvaceae and distributed in all parts of the tropical and sub-tropical region of India. All parts of the plant have been recognized to have medicinal properties. The plant is commonly called Thutti in Tamil. The traditionally, the plant used as anthelmintic, anti-inflammatory, and is useful in urinary and uterine discharges, piles, and lumbago [23], jaundice, ulcer, and leprosy. *A. indicum* leaves are used in the treatment of toothache, lumbago, piles, anti-fertility, and liver disorders [24]. Root and bark are used as aphrodisiac, antidiabetic [25], nervine tonics and diuretic. The plant extracts and their products for antimicrobial activity have shown that a potential source of novel antibiotic prototypes of higher plants [26]. *A. indicum* leaves are used in the treatment of toothache, lumbago, antifertility, and liver disorders [27]. Bark and root are used as antidiabetic, aphrodisiac [28] nervine tonic, and diuretic. The plant extracts and their products for antimicrobial activity have shown that a potential source of novel antibiotic prototypes of higher plants [29].

Thus, this invasination is carried out on management of disease by aqueous extracts of *Abutilon indicum, Prosopis juliflora,* and *Acacia arabica* plants.

4.2   MATERIALS AND METHODS

4.2.1   PLANT MATERIAL

Commonly available weed plants were collected in the surrounding areas of Latur district, Maharashtra. The plants used for this study are *Abutilon indicum, Prosopis juliflora,* and *Acacia arabica.*

4.2.2   PREPARATION OF PLANT EXTRACTS

Plant leaves were washed with tap water followed by sterile distilled water and then air-dried at Room temperature. Dried leaves were then powdered. 50 grams of powdered material was percolated with 250 ml water for 72 hours. The percolate was mixed thoroughly for every 12 hours. Percolate was filtered through Whatman no.1 filter paper. The filtrate was concentrated at 35°C and stored at 4°C till the use.

4.2.3   FIELD VISIT AND COLLECTION OF BACTERIAL SAMPLES

The field visit was undertaken in major pomegranate growing regions of Latur, Maharashtra, India, i.e., Murud, and Harangul during the month of June to August 2018. During the field survey, the randomly selected plant parts were inspected at the fields of for the incidence of bacterial blight. Distribution of bacterial blight of pomegranate was recorded in these areas. Plants were diagnosed as infected based on typical symptoms of bacterial blight. The symptoms were yellow water-soaked lesions at early stages and corky, dark oily spots at later stages of infection. The suspected plant leaves and fruits (Figure 4.1) were collected and transferred into sterilized plastic bags and brought to Research laboratory of Microbiology Department of Dayanand Science College, Latur for the further studies.

(A) (B)

(C) (D)

FIGURE 4.1 Bacterial blight symptoms on pomegranate fruits and leaves. (A) Bacterial blight symptoms on pomegranate fruit; (B) bacterial blight symptoms on fruit; (C) bacterial blight symptoms on pomegranate fruit; (D) bacterial blight symptoms on fruit and leaves.

4.2.4   STORAGE OF SAMPLE

The samples were kept in polythene bag and stored at 40°C in refrigerator till the use.

4.2.5   ISOLATION OF THE BACTERIA

The bacteria were isolated from the infected leaves, and fruits of pomegranate collected from different habitats of Latur district. These samples were washed, air dried, and then disinfected with 0.1% HgCl$_2$ for about 30–60 seconds and washed thoroughly with sterile water to remove traces of HgCl$_2$. The selected spots were cut with sterilized blade in a sterile Petri dish containing few drops of sterile distilled water in order to allow the bacteria to diffuse out. A loop full of suspension was then transferred with the help of sterilized bacteriological needle to sterilized Petri plates filled with nutrient agar medium with (NA) and incubated at 28°C for 24–72 hr. After 2–3 days, incubated plates were observed for the presence of typical pale yellow, glistening colonies (Figure 4.2) which were transferred to the NA slants and maintained in laboratory conditions for further studies.

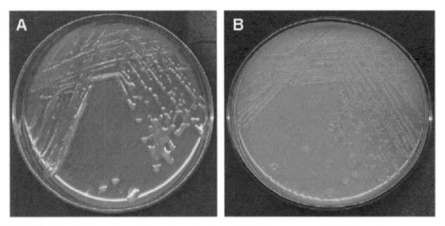

FIGURE 4.2   Pale yellow, glistening colonies of Xanthomonas on the NA medium.

4.2.6   IDENTIFICATION AND CONFIRMATION OF ISOLATE

Identification and Confirmation of *Xanthomonas axonopodis pv. Punicae* was done by using the following methods.

4.2.6.1   MORPHOLOGICAL CHARACTERS

The morphological characters such as shape, gram reaction, and pigmentation characters were studied as described by the Society of American Bacteriologists, Bradbury [30], and Schaad and Stall [31] (Table 4.1).

TABLE 4.1 Cultural and Morphological Characteristics of *Xanthomonas Axonopodis pv. Punicae* Isolate on Nutrient Agar Media

Sl. No.	Colony Characters	*Xanthomonas Axonopodis pv. Punicae*
1.	Color	Yellowish
2.	Size of the colony	Medium to large
3.	Shape of the colony	Small Circular colonies
4.	Cell shape	Single rod
5.	Appearance	Slightly raised, glistening
6.	Elevation	Convex
7.	Margin	Entire margin
8.	Texture	Highly mucoid

4.2.6.2 GROWTH RATE AT 28°C AND 37°C

The effect of varied temperature levels on the growth of *Xanthomonas axonopodis pv. punicae* was studied and data so obtained is presented in Table 4.2. Isolate of *Xanthomonas axonopodis pv. punicae* were tested at temperature 28°C and 37°C on Nutrient agar medium. The data clearly indicated that the temperature of 28°C was found to be optimum for the growth of the pathogen due to appearance of relatively maximum number of colonies at this temperature. Isolate grew well at 28°C but no growth was observed at temperature of 37°C. Growth of *Xanthomonas axonopodis pv. punicae* started 72 hrs after incubation. Maximum growth was observed after 120 hrs of incubation at 28°C but no growth was observed after 48 hrs of incubation. At 37°C, no growth was observed up to 120 hrs. Similar work on temperature requirement was carried out by Hingorani and Mehta [5]. They found that the pomegranate bacterium grows well at a cordial temperature of 30°C and can tolerate a minimum and maximum temperature of 5°C and 40°C, respectively. Gour et al., [32] also got the similar results while working with *X. axonopodis* pv. *vignicola*, the causal agent of leaf blight of cowpea. They have recorded the maximum growth of the pathogen at a temperature level of 30°C, whereas, Manjula [33] recorded the highest number of colonies of *Xap.* at a temperature of 27°C.

4.3 BIOCHEMICAL VARIABILITY

Methodology followed for the following experiments on biochemical variability is according to Schaad [31].

TABLE 4.2   Growth Rate of *Xanthomonas Axonopodis pv. Punicae* Isolate

Sl. No.	Isolate	Growth Rate							
		28°C				37°C			
		48 hrs	72 hrs	96 hrs	120 hrs	48 hrs	72 hrs	96 hrs	120 hrs
1.	*Xanthomonas axonopodis pv. punicae*	+	++	+++	++++	-	–	–	–

Colony growth: '–': no growth, '+': less growth, '++': moderate growth, '+++': maximum growth).

4.3.1   LACTOSE UTILIZATION

Carbon source (lactose) was filter sterilized and mixed with autoclaved, cooled Dye's medium along with 1.2% purified agar. The pH was adjusted to 7.2. Bacterial isolate were spot inoculated with replica plating method and incubated at 30°C for 3, 7 and 14 days. Growth was compared with control, where carbon source was not supplemented [31].

4.3.2   STARCH HYDROLYSIS

Medium used for Starch hydrolysis (see Appendix) was sterilized by auto-claving and poured into sterilized Petri plates. These plates were inoculated and incubated at 30°C for 7 days. The plates were flooded with Lugol's iodine and allowed to act for few mins. The presence of starch hydrolysis was indicated by the presence of clear zones and *vice-versa*. The zone hydrolyzed was measured for each isolate.

4.3.3   ACID PRODUCTION FROM SUCROSE, MALTOSE, DEXTROSE

The acid production by the isolate of *Xanthomonas axonopodis pv. punicae* was tested by using Peptone water medium of Dye. About 10 ml of medium was dispensed in each test tube. This medium was sterilized in an autoclave for 15 min. To these tubes, filter sterilized carbohydrates *viz.*, Sucrose, Maltose, and Dextrose were added at 0.14% concentration. The tubes were inoculated with 0.1 ml of 24 hrs old bacterial culture and incubated at room temperature for three days. Change in the color of the medium confirmed the acid production (Figure 4.3).

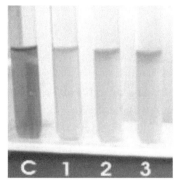

FIGURE 4.3   Acid production from sucrose, maltose, dextrose. (C: Control, 1: Sucrose, 2: Maltose, 3: Dextrose).

4.3.4   TESTING ANTIBACTERIAL ACTIVITY

After conformation of *Xanthomonas axonopodis pv. punicae,* the antibacterial activity of plant extract was preliminarily screened by well diffusion assay. LB agar plates were spread plated with 20 µL of bacterial strain (1 × 10^8 CFU/ml). The wells of 6 mm diameter were made in the agar plates. Each plant extract was tested for antibacterial activity by adding 40 µL of extracts in different concentrations *viz.*, 50, 100, and 200 mg ml^{-1}. The experiment was repeated thrice. The plates were incubated at 37°C for 24 hours. Subsequently, the plates were examined for zone of inhibition (ZOI) and diameter was measured in mm after subtracting well diameter. To determine Minimum inhibitory concentration (MIC), required quantity of extracts were added into the LB broth of 4 ml to bring initial concentration of 20 mg ml^{-1}. In each test tube, 0.1 ml of standardized inoculum (1 × 10^8 CFU/ml) was added. Two control tubes were maintained for each test batch namely extract control (tube containing plant extract and LB medium without inoculum) and organism control (tube containing LB medium and inoculum). The test tubes were incubated at 37°C for 24 hours. The lowest concentration (highest dilution) of plant extract that produced no visible growth (no turbidity) recorded as MIC. Minimum bactericidal concentration (MBC) was assayed by sub-culturing test dilutions on a drug-free solid medium. The plates were incubated for 24 hours at 37°C. The lowest concentration of the antimicrobial at which no single colony observed after sub-culturing is regarded as MBC [34].

4.4 RESULTS

4.4.1 WELL DIFFUSION ASSAY

The preliminary screening of selected three plant extracts against the *Xanthomonas axonopodis pv. punicae* was done using well diffusion method. The ZOI greater than 5 mm diameter is found to be having significant activity against particular bacteria [35]. The aqueous extracts were sensitive against *Xanthomonas axonopodis pv. punicae* tested at different concentrations (Table 4.1).

The extract of *Acacia aarabica* was sensitive against *Xanthomonas axonopodis pv. punicae* tested at different concentrations. The same extract was much effective at low concentration of 50 mg ml^{-1}, i.e., 11 mm ZOI. *P. juliflora* extract arrested the growth of *Xanthomonas axonopodis pv. punicae* (24 mm) at 200 mg ml^{-1}. The *P. juliflora* extract shows a ZOI about 21 mm and 19 mm at 100 and 50 mg ml^{-1} respectively. Based on ZOI, Satish et al. reported significant antibacterial activity *of P. juliflora* and *A. arabica* against *X. campestris pathovars*. The extracts of *A. indicum was* exhibited the antibacterial activity of (11 mm) ZOI at 200 mg ml^{-1} (Table 4.3).

TABLE 4.3 Antibacterial Activity of Plant Extracts Showing Zone of Inhibition

Sl. No.	Plant Species	Conc. mg ml^{-1}	Zone of Inhibition (mm)*
1.	*Acacia arabica*	50	11
		100	15
		200	16
2.	*Prosopis juliflora*	50	19
		100	21
		200	24
3.	*Abutilon indicum*	50	6
		100	9
		200	11

4.4.2 MINIMUM INHIBITORY AND BACTERICIDAL CONCENTRATION

The antibacterial activity of *A. arabica* and *P. juliflora* extracts are found to be high against all the bacteria. The maximum activity recorded in *P. juliflora* (MIC = 1.03 mg ml^{-1} and MBC = 0.15 mg ml^{-1}) and *A. arabica* (MIC = 1.00372 mg ml^{-1} and MBC = 2.58 mg ml^{-1}) against *X. axonopodis pv. Punicae*. Raghavendra et al. [11] reported significant antibacterial activity of *Acacia nilotica* extracts against *Xanthomonas pathovars* and human pathogenic bacteria tested, while the lowest activity was recorded by *A. indicum* (MIC = 0.619 mg ml^{-1} and MBC = 0.923 mg ml^{-1}). These findings indicate that the extracts of *Prosopis juliflora* and *Acacia arabica* are potential to use in the management of plant diseases. Further phytochemical analysis is required to identify the active components of plant extracts showing antimicrobial activity (Table 4.2 and Figure 4.4).

TABLE 4.4 MIC and MBC of Plant Extracts Against *Xanthomonas axonopodis pv. Punicae*

Sl. No.	Plant Species	MIC (mg ml^{-1})	MBC (mg ml^{-1})
1.	*Acacia arabica*	1.00372	2.58
2.	*Prosopis juliflora*	1.03	0.15
3.	*Abutilon indicum*	0.619	0.923

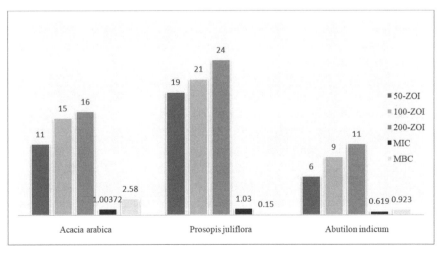

FIGURE 4.4 Graphical representation of susceptibility of plant extracts.

4.5   DISCUSSION

The antibacterial activity of extracts of *Abutilon indicum, Prosopis juliflora,* and *Acacia arabica* plants may be due to the presence of phytochemicals. There are several evidences on the presence of antimicrobial metabolites like tannins, flavonoid, glycosides, essential oils, furostanol, spirostanol, saponins, phytosterols, amides, alkaloids, etc., in the studied plant species [36]. Secondary metabolites such as polyphenols are not required for plant development and growth, but are involved in plant communication and defense [37]. Tannins and saponins are plant metabolites well known for their antimicrobial properties [38]. Flavonoids have both antifungal and antibacterial activities. They possessed anti-inflammatory properties also [39]. Saponins, flavonoids, terpenes, and steroids are known to have antimicrobial and curative properties against several pathogens [40]. The antimicrobial effects of plant materials are believed due to secondary products present in the plant, although it is usually not attributed to a single compound, but to a combination of metabolites [41]. The exact mechanism of bacterial growth inhibition of the plant materials is yet to know. One of the most important mechanisms is the hydrophobic activity of the bioactive compounds which enables them to partition the lipids of the bacterial cell membrane and mitochondria, disturbing the cell structures and rendering them more permeable. Extensive leakage from the bacterial cells or the exit of critical molecules and ions will consequences death of the bacteria [42].

4.6   CONCLUSION

Many synthetic antibiotics are used to control several phytopathogens. The increased awareness of environmental problems with these chemical antibiotics has led to the search for non-conventional chemicals of biological origin for the management of these diseases. Bactericides of plant origin can be one approach to disease management because of their eco-friendly nature [12]. The products of plant origin are of greater advantage to user, the public, and radical environmentalists. Laboratory screening of plant extracts has given encouraging results, indicating their potential use in the management of disease caused by *Xanthomonas* species. Plant extracts resulted in antibacterial activity is potential to use in the management of plant diseases as an alternative to chemical antibiotics.

Further phytochemical analysis is required to identify the bioactive compounds responsible for antibacterial activity. Yet the results gained by these methods provide simple data that make it possible to classify extracts in respect to their antioxidant potential. As can be observed also from the present data, antibiotic activity does not necessarily correlate with high amounts of phytochemicals, and that is why both phytochemicals content and antibiotic activity information must be discussed when evaluating the phytochemical potential of extracts. On the basis of this study, the most potent Finnish plant sources for natural antibiotic agents are medicinal plants and vegetable peels, and different tree materials. Further work is underway to confirm the antibiotic effect of these promising plant extracts by using other types of bacterial strains and to characterize the active phytochemicals, their mechanism of action, and their possible interactive effects together with harmful effects to the environment.

Since extracts of *Abutilon indicum, Prosopis juliflora,* and *Acacia arabica* have shown the potential antimicrobial activity against all the bacterial fungal strains tested. Moreover, the present investigation indicated the potential source of antimicrobial activity of leaf extracts of *Abutilon indicum, Prosopis juliflora* and *Acacia arabica* against a variety of biologically active organisms and it is hoped that the present results will provide a starting point for investigations aimed at exploiting new natural antibacterial substances present in the *Abutilon indicum, Prosopis juliflora,* and *Acacia arabica.*

ACKNOWLEDGMENT

The author wants to thanks Dr. Jaiprakash Dargad (Principal, Dayanand Science College, Latur) for providing Research Laboratories and Chemicals to work. The author wants to extend her thanks to the Dayanand Education Society for providing Research funding to the author. Special Thanks to Dr. R. S. Shinde for his impressive collaboration with the author. Also, the author heartily acknowledges to all supporting teachers and laboratory assistants of Dayanand Science College, Latur.

KEYWORDS

- *Acacia arabica*
- **minimum bactericidal concentration**
- **minimum inhibitory concentration**

- *Prosopis juliflora*
- **synthetic antibiotics**
- **zone of inhibition**

REFERENCES

1. Malhotra, N. K., Khajuria, H. N., & Jawanda, (1983). Studies on physio-Chemical characters of pomegranate cultivars II. Chemical characters. *Punjab Horti. J., 23*, 158.
2. Bose, T. K., & Pomegranate, (1985). In: *Fruit of India-Tropical and Subtropical* (p. 637). B. Mitra Naya Prakash Publications, Calcutta.
3. Srivastava, & Umesh, C., (2008). Horticulture in India: A success story. *Crop Care, 34*, 23–32.
4. Raghavan, R., (2007). *Oily Spot of Pomegranate in India (Maharashtra).*Express India.
5. Hingorani, M. K., & Mehta, P. P., (1952). Bacterial leaf spot of pomegranate. *Indian Phytopath., 5*, 55–56.
6. Vauterin, L., Haste, B., Kersters, K., & Swings, J., (1995). Reclassification of *Xanthomonas. Int. J. Syst. Bacterial., 45*, 475–489.
7. Hingorani, M. K., & Singh, N. J., (1959). *Xanthomonas punicae* sp. Nov. on pomegranate (*Punica granatum* L.). *Indian J. Agric. Sci., 29*, 45–48.
8. Petersen, Y., Mansvelt, E. L., Venter, E., & Langenhoven, W. E., (2010). Detection of *Xanthomonas axonopodis pv. Punicae*causing bacterial blight on pomegranate in South Africa. *Australasian Plant Pathology, 39*, 544–546.
9. Chand, R., & Kishun, R., (1991). Studies on bacterial blight (*Xanthomonas campestris* pv. *punicae*) on pomegranate. *Indian Phytopathology, 44*, 370–372.
10. Rani, U., Verma, K. S., & Sharma, K. K., (2001). Pathogenic potential of *Xanthomonas axonopodis pv. Punicae* and field response of different pomegranate cultivars. *Plant Disease Research, 16*, 198–202.
11. Raghavendra, M. P., Satish, S., & Raveesha, K. A., (2006). *In vitro* evaluation of antibacterial spectrum and phytochemical analysis of *Acacia nilotica. J. Agrl. Tech., 2*, 77–88.
12. Bolkan, H. A., & Reinert, W. R., (1994). Developing and implementing IPM strategies to assist farmers, an industry approach. *Plant Dis., 78*, 545–550.
13. Mattana, C. M., Satorres, S. E., Escobar, F., Sabini, C., Sabini, L., Fusco, M., & Alcaraz, E., (2012). Antibacterial and cytotoxic activities of *Acacia aroma* extracts. *Emir. J. Food Agric., 24*(4), 308–313.
14. Negi, B. S., & Dave, B. P., (2010). *In vitro* antimicrobial activity of *Acacia catechu* and its phytochemical analysis. *Indian J. Microbiol., 50*(4), 369–374.
15. Kim, H. S., Han, S. S., Oh, K. W., Jeong, T. S., & Nam, K. Y., (1987). Effects of ginseng saponin on antimicrobial activities of some antibiotics. *Korean J. Mycol., 15*, 87–91.
16. Seigler, D. S., (2003). Phytochemistry of *Acacia sensu*lato. Biochemical systematics and ecology. *Biochem. Sys. Ecol., 31*(8), 845–873.
17. Yasir, M., Jain, P., Debajyoti, & Kharya, M. D., (2010). Hypoglycemic and antihyperglycemic effect of different extracts of *Acacia arabica lamk* bark in normal and alloxan induced diabetic rats. *Int. J. Phytomed., 2*, 133–138.

18. Pasiecznik, N., Felker, P. J., Harris, L. N., Harsh, G., Cruz, J. C., Tewari, K. C., & Maldonado, L. J., (2001). *The Prosopis Juliflora- Prosopis Pallida Complex: A Monograph.* HDRA, Coventy, UK.

19. Choge, S. K., Ngunjiri, F. D., Kuria, M. N., Busaka, E. A., & Muthondeki, J. K., (2002). *The Status and Impact of Prosopis spp.* In Kenya, Unpublished report, Kenya Forestry Research Institute and Forest Department.

20. Singh, S., Swapnil, & Verma, S. K., (2011). Antibacterial properties of alkaloid rich fractions obtained from various parts of *Prosopis juliflora. Int. J. Pharma. Sci. Res., 2*(3), 114–120.

21. Pasiecznik, N. M., Harris, P. J. C., & Smith, S. J., (2004). *Identifying Tropical Prosopis Species: A Field Guide.* Coventry: HDRA Publishing.

22. Burkart, A., (1976). A monograph of the genus *Prosopis (Leguminosae subfam.* Mimosoideae). *J Arnold Arbor., 57,* 219–249.

23. Porchezhian, E., & Ansari, S. H., (2000). *Pharmazie., 55,* 702–703.

24. Anyensu, E. S., (1978).*Medicinal Plants of West Africa* (p. 110). Algonac, Michigan. Publications, Inc.

25. Lakshmayya, Nelluri, N. R., Kumar, P., Agarwal, N. K., Gouda, T. S., & Setty, S. R., (2003). *Indian Journal of Traditional Knowledge, 2*(1), 79–83.

26. Afolayan, A. J., (2003). *Pharmaceutical Biology, 41,* 22–25.

27. Anyensu, E. S., (1978). *Medicinal Plants of West Africa* (p. 110). Algonac, Michigan. Publications Inc.

28. Lakshmayya, Nelluri, N. R., Pramod, K., Agarwal, N. K. T., Shivaraj, G., & Ramachandra, S. S., (2003). Phytochemical and pharmacological evaluation of leaves of *Abutilon indicum. Indian Journal of Traditional Knowledge, 2,* 79–83.

29. Afolayan, A. J., (2003). Extracts from the shoots of arctotisartotoides inhibit the growth of bacteria and fungi. *Pharmaceutical Biology, 41,* 22–25.

30. Bradbury, J. F., (1970). Isolation and preliminary study of bacteria from plants. *Rev. Pl. Pathol., 49,* 213–218.

31. Schaad, N. W., & Stall, R. E., (1988). *Xanthomonas. In: Laboratory Guide for Identification of Plant Pathogenic Bacteria* (2ndedn., pp. 81–94).

32. Gour, H. N., Ashiya, J., Mali, B. L., & Ranjan, N., (2000). Influence of temperature and pH on the growth and toxin production by *Xanthomonas axonopodis pv. Vignicola* inciting leaf blight of cowpea. *J. Mycol. Pl. Path., 30,* 389–392.

33. Manjula, C. P., (2002). *Studies on Bacterial Blight of Pomegranate (Punica granatum L.) Caused by Xanthomonas Axonopodis pv. Punicae* (p. 98)*.* MSc (Agri.) Thesis submitted to Univ. of Agri. Sci., Bangalore.

34. 34. Akinyemi, K. O., Oladapo, O., Okwara, C. E., Ibe, C. C., & Fasure, K. A., (2005). Screening of crude extracts of six medicinal plants used in South West Nigerian unorthodox medicine for anti-methicillin resistant *Staphylococcus aureus* activity. *BMC Complement Altern. Med.,* 6.

35. Palombo, E. A., & Semple, S. J., (2001). Antibacterial activity of traditional Australian medicinal plants. *J. Ethanopharmacology, 77,* 151–157.

36. Ganjewala, D., Sam, S., & Khan, K. H., (2009). *Eurasia Journal of Biological Sciences, 3*(10), 69–77.

37. Parekh, J., Nair, R., & Chanda, S., (2005).*Indian J. Pharmacol. (India) (In Press).*

38. Trechesche, R., (1971). Advances in chemistry of antibiotic substance from higher plant. *Pharmacognosy and Phytochemistry Proceeding of the 1st International Congress* (pp. 274–276). Verlag, Berlin, Heidelberg, New York.

39. Ogundaini, A. O., (2005). *Form Gene into Medicine: Taking a Lead form Nature* (pp.12–15). An Inaugural lecture delivered at Oduduwa Hall, Obafemi Awolowo University, Ile-Ife, Nigeria.

40. Usman, H., Abdulraliman, F., & Osuji, J. E., (2007). *Afr. J. Trad. Compl. Alten. Med., 6,* 476–480.

41. Hui-Mei, L., Hsien–Chun, T., Chau-Jong, W., Jin-Jin, L., Chia-Wen, L., & Fen-Pi, C., (2008). *Chem. Biol. Interact., 171,* 283–293.

42. Fawzi, M. M., Anwar, H. S., Ameenah, G., & Muhammad, I. C., (2012). *BMC Com Alt Med., 12,* 165.

CHAPTER 5

Microbial Biotechnology: Synthesis, Production, Challenges, and Opportunities

AKSHADA A. BAKLIWAL, SWAPNALI A. PATIL,
MEGHAWATI R. BADWAR, and SWATI G. TALELE

Department of Pharmaceutics, Sandip Institute of Pharmaceutical Sciences, Nashik Maharashtra India,
E-mail: swatitalele77@gmail.com (S. G. Talele)

5.1 INTRODUCTION

Biotechnology might be characterized as 'the usage of living life forms in frameworks or procedures for the creation of important items; it might include microscopic organisms, parasites, yeast, cells of higher plants and creatures or subsystems of any of these or detached parts from living issue' [6].

1. **Parts of Biotechnology:** The meaning of biotechnology can be additionally separated into various territories known as red, green-blue, and white.

   - **Red Biotechnology:** This region incorporates restorative methods, for example, using living beings for the generation of novel medications or utilizing immature microorganisms to supplant/recover harmed tissues and perhaps recover entire organs. It could basically be called restorative biotechnology.
   - **Green Biotechnology:** It applies to agribusiness and includes such forms as the advancement of nuisance safe grains and the quickened development of illness safe creatures.

- **Blue Biotechnology:** Once in a while referenced, incorporates forms in the marine and amphibian conditions, for example, controlling the multiplication of toxic water-borne living beings.
- **White Biotechnology:** It includes mechanical procedures, for example, the creation of new synthetic compounds or the advancement of new fills for vehicles.

Microbial biotechnology includes the abuse, hereditary control, and modifications of miniaturized scale life forms to make business important items, and that additionally includes aging and different upstream and down-stream forms [7]. Microorganisms produce a stunning exhibit of important items, for example, macromolecules (for example proteins, nucleic acids, starch polymers, even cells) or littler particles and are normally partitioned into metabolites that are fundamental for vegetative development (essential metabolites) and those which give points of interest over antagonistic condition (auxiliary metabolites). They generally produce these mixes in modest quantities that are required for their very own advantage [8, 9].

5.1.1 REQUIREMENT FOR EFFECTIVE BIO-PRODUCTION FRAMEWORKS FOR NUTRIENT, BIO-PIGMENTS, ANTIOXIDANTS, AND RELATED WELLBEING FACTORS

Vitamins and related compounds, such as pro-vitamins, bio-pigments, and antioxidants, have a place with those couple of synthetic compounds that intrigue in a constructive manner to a great many people. These terms sound synonymous with imperativeness, great wellbeing, and mental quality, even to the layman. All of us need his/her every day admission of (genius) nutrients and cancer prevention agents, ordinarily given by a decent and differed diet. In any case, current nourishment propensities or inclinations, sustenance availabilities, just as nourishment handling, safeguarding or cooking philosophies and advancements, don't generally guarantee an adequate adjusted normal day by day (pro)vitamin supply to a solid individual, and significantly more so for a focused or wiped out person. Today, present-day society is rarely stood up to with the famous avitaminosis of the past, understood toward the Western World, yet they do in any case happen every now and again in overpopulated, poverty or famine-struck areas on our globe, just as for shockingly enormous populace bunches in created nations. Aside from their in vivo nutritional-physiological jobs as basic development components and coenzymes for man, creatures, plants, and microorganisms, nutrients, and related mixes are progressively being presented as nourishment and as feed added substances, as medical-therapeutically operators, as wellbeing advancing guides. These days an amazing number of handled nourishments, encourages, makeup, pharmaceutical, and concoction plans contain extra-added (master) nutrients or vitamin-related mixes, and single and multivitamin arrangements are normally taken or endorsed.

Notwithstanding their notable dietary, physiological, and therapeutic significance, (pro)vitamins and related wellbeing mixes have likewise found large-scale specialized applications, for instance, as cancer prevention agents (D-isoascorbic corrosive as the C5-epimer of nutrient C, glutathione or GSH, tocopherol or nutrient E, carotenoids, wine, and tea polyphenols), as acidulants (ascorbic corrosive or nutrient C) and as bio-pigments (yellow-orange-red carotenoids, yellow riboflavin or nutrient B2, red Monascus-pigments) in the nourishment, feed, restorative, concoction, nutraceutical, and pharmaceutical segments. There is particularly a requirement for normal shades of (miniaturized scale) natural starting point to supplant manufactured shades and colorants. Certain contagious carotenoids (*Blakeslea trispora*-beta-carotene, *Xanthophyllomyces dendrorhous*-asthaxanthin) and algal carotenoids (*Dunaliella salina*-xanthophylls, lycopene), cochineal-carminic corrosive from scale bugs, blue-purple phycocyanin from Arthrospira-cyanobacteria, and parasitic dim rosy monascin-pigments are as of now utilized in this regard, yet these bioprocesses should be additionally improved as to yield and biotechnology devices included. The above contemplations point towards an additional requirement for (bio)synthesis and supply of (pro)vitamin, biopigment, cancer prevention agent, and related wellbeing atoms, over the level, gave normally from the microbial, plant, and creature sustenance sources.

5.1.2   CHANGE FROM EXTRACTION INNOVATION OVER CHEMICAL SYNTHESIS TOWARDS INDUSTRIAL BIOTECHNOLOGY-BASED FORMS

Till a couple of decades, back most included nutrients and related wellbeing mixes were to be sure mechanically arranged by means of extraction advancements. Thinks or concentrates got from vitamin-rich or hued characteristic staple nourishment items (of plant, creature, or microbial source), be that as it may, discover now generally little use in the sustenance, feed, pharmaceutical or corrective division. Aside from their high value, a portion of the reasons are:

- The level of nutrients and related wellbeing mixes in the normal plant/ creature source is generally moderately low and vacillates radically.
- Their organoleptic introduction and shelf-life are regularly not ideal.
- Water-soluble nutrients are effectively lost by fluid extraction or different controls of these regular sustenance nutrient sources.
- (Pro)nutrients and related wellbeing mixes are labile particles during the procedure of collect, conservation, stockpiling (or during the planning of staples) and are commonly delicate to pH, heat (riboflavin or B2, B5, pyridoxine or B6, folic corrosive or B9, nutrient C, nutrient E), light (B2, B6, B9, nutrient B12, C, nutrient D), oxygen (B9, C, D, fundamental unsaturated fats or EFA's).

These disadvantages have prompted the mechanical assembling of most nutrients and related factors by a substance or microbial combination course. As of now, a few nutrients are made artificially (pro-vitamin A, cholecalciferol or D3, E, nutrient K1 or phylloquinone and thiamine or B1, B5, B6, D-biotin or B7, B9), albeit enzymatic, microbiological, and additionally biotechnological strategies develop or exist, in spite of the fact that not monetarily gainful so far. For a portion of these particles or their forerunners, biotechnological procedures are being created, albeit for sure not aggressive up 'til now with substance amalgamation.

Two biotechnological courses, coordinated aging procedures, and biocatalysis take step by step over from synthetic union for the majority of these artificially perplexing atoms. The two advances were at first regularly protected just when substance procedures neglected to be fruitful or were uneconomical. These days they become regularly first decision advances for a few reasons: they depend on inexhaustible assets, convey basic just as unpredictable particles legitimately in an attractive chiral structure and in a financially ideal manner and they are considered in the public eye as perfect, feasible, and reuse advances.

5.2   HISTORY

Since 5000 BC, organisms have been utilized for making wine, vinegar, curd, raised bread, and so on such forms that depend on normal abilities of microorganisms which are regularly alluded to as old biotechnology. Biotechnology, in the twentieth century, united industry and horticulture [1, 2].

Hungarian designer Karl Ereky first authored the term 'biotechnology' in 1919, which means the generation of items from crude materials with the guide of living creatures. The creation advances and procedures engaged with creature cultivation, farming, agriculture, and so on, use plants and creatures to deliver valuable items. Be that as it may, such innovations are not viewed as biotechnology since they are for some time perceived and well-established disciplines in their own right. Today, the misuse of the creature and plant cells refined in vitro just as their constituents for producing items/administrations is a vital part of biotechnology [3–5].

5.3   SYNTHESIS AND PRODUCTION OF MICROBIAL ENZYMES

Microorganisms and parasites produce most mechanical catalysts. Normally happening microorganisms are the most gainful makers of compounds. This information has been abused by the industry for over 50 years. Microbes and parasites are the microorganisms most appropriate to the modern generation of proteins. They are anything but difficult to deal with, can be developed in enormous tanks without light, and have a high development rate [10].

The perfect microorganism develops rapidly and creates heaps of the ideal compounds at gentle temperatures while expending cheap supplements. Be that as it may, as most things throughout everyday life, the perfect microorganism is rare. Most microorganisms found in the wild are not appropriate to taming in enormous aging tanks. Some lone produce little amounts of protein or set aside a long effort to develop. Others can create undesired results that would exasperate mechanical procedures. So for the modern generation, an ideal microorganism is the first prerequisite [11].

5.3.1   ENZYME PRODUCTION IN INDUSTRIES

Various microorganisms have been utilized for modern protein creation, fluctuating from eukaryotic frameworks, for example, yeast, and organisms, to prokaryotic framework including both Gram-positive and Gram-negative microscopic organisms.

The fundamental component of catalyst union incorporates interpretation, interpretation, and post-translational preparing which is exceptionally rationed. Notwithstanding, a few contrasts exist between different classes of creatures, just as some essential contrasts among prokaryotic and eukaryotic life forms. The catalysts themselves vary in their atomic structure, number of polypeptide chain, level of glycosylation, and isoelectric point. Albeit every one of the distinctions impacts the manufactured example, the fundamental protein union instruments are comparable enough to permit a general treatment of the microbiological creation process. In any case, contrasts exist between the generation energy of various catalysts by various microorganisms due to their shifted physical attributes and development design, which requires the enhancement of every creation procedure independently [12, 13].

5.3.2   INDUSTRIAL ENZYME PRODUCTION TECHNOLOGY

Fermentation technology have been utilized only for the creation of mechanical proteins, ideally by microorganisms, for example, microbes or growths under deliberately controlled conditions because of their simplicity of augmentation and dealing with. Microorganisms utilized are GRAS (generally recognized as safe) strains because of their application in the nourishment and feed industries.

Fermentation procedure configuration is interdisciplinary and requires learning of both substance building and microbial physiology to be effectively scaled up. Submerged fermentation (SmF) and solid-state fermentation (SSF) are the two significant fermentation advances accessible. Both of these advancements offer a few advantages and have their very own confinements. Most enterprises utilize SmF for compound generation; be that as it may, there has been resurgence in fame of SSF for couple of utilizations and explicit ventures [14, 15].

5.3.2.1   SUBMERGED FERMENTATION

Fermentation did within the sight of an overabundance of free water is named as submerged fermentation. The utilization of an oxygen consuming submerged culture in a blended tank reactor is a run of the mill mechanical procedure for compound creation including microorganisms that produce an extracellular protein. It is the favored innovation for mechanical compound generation because of the simplicity of dealing with at an enormous scale when contrasted with SSF. Huge scale fermenters for SmF, changing in volume from thousands to a hundred thousand liters, are all around created and offer online command more than a few parameters, for example, pH, temperature, DO (dis-fathomed oxygen), and froth development; and additionally there is no issue of mass exchange and warmth evacuation. Hence, these are a portion of the advantages which make this creation innovation better than SSF and broadly acknowledged for mechanical metabolite generation. The medium in the SmF is fluid which stays in con-affability with the microorganisms. A supply of oxygen is basic in the SmF, which is done through a sparger. Stirrers and impellers assume a significant job in these fermenters for blending gas, biomass, and suspended particles [16].

There are four primary methods for developing the microorganisms in the SmF. These are batch culture, fed-batch culture, perfusion batch culture and continuous culture. In the batch culture, the microorganisms are vaccinated in fixed volume of medium. On account of fed-batch culture, the concentrated segments of the supplement are progressively added to the batch culture. In the perfusion batch culture, the expansion of the way of life and withdrawal of an equivalent volume of utilized without cell medium is performed. In the continuous culture, new medium is included into the group framework at the exponential period of the microbial development with a relating withdrawal of the medium containing the item. The continuous development gives a close adjusted development, with little vacillation of the supplements, metabolites, cell numbers, or biomass [17].

By and by scale-up impacts are more articulated for the oxygen-consuming procedure than the anaerobic procedure. To accomplish air circulation, the tumult is kept up during the scale-up in the fermenter to keep up consistent oxygen supply. Scale-up entanglements emerge from cell reactions to appropriated estimations of broke down oxygen, temperature, pH, and supplements. For catalyst generation, the economy of scale prompts the utilization of fermenters with a volume of 20–200 m^3. The accompanying issues of mass and warmth move are typically dismissed in little fermenters and at low cell densities. Be that as it may, in modern microbiology, with the previously mentioned fermenter volumes and the financial need of utilizing the most elevated conceivable cell densities, transport procedures must be considered. These can restrain the metabolic rates, for instance, oxygen constraint drives the microorganisms to react with changes in their physiological example. In these conditions, the ideal control of microbial digestion is lost. In the controlled activity of a mechanical procedure, metabolic rates must be restricted to a level just beneath the vehicle limit of the fermenter. Thusly, the most elevated conceivable profitability in a fermenter is acquired at maximal vehicle limit [18].

5.3.2.2   SOLID STATE FERMENTATION (SSF)

Current improvements in biotechnology are yielding new applications for chemicals. SSF holds enormous potential for the creation of catalysts. It very well may be of exceptional enthusiasm for those procedures where the unrefined aged items might be utilized legitimately as protein sources. This framework offers various points of interest over the SmF framework, including high titer, moderately higher grouping of the items, less emanating age, the necessity for straightforward maturation hardware, less prepared work, and so forth [19].

Countless microorganisms, including microbes, yeast, and parasites, produce various gatherings of catalysts. Determination of a specific strain, notwithstanding, stays a monotonous undertaking, particularly when industrially equipped compound yields are to be accomplished.

Agro-industrial residues are commonly viewed as the best substrates for the SSF procedures, and utilization of SSF for the generation of catalysts is no special case to that. Various such substrates have been utilized for the development of microorganisms to deliver chemicals. A portion of the substrates that have been utilized incorporate sugar stick bagasse, wheat grain, rice grain, maize grain, gram grain, wheat straw, rice straw, rice husk, soy frame, sago hampas, grapevine trimmings dust, saw dust, corncobs, coconut coir essence, banana squander, tea squander, cassava squander, palm oil factory squander, aspen mash, sugar beet mash, sweet sorghum mash, apple pomace, shelled nut supper, rapeseed cake, coconut oil cake, mustard oil cake, cassava flour, wheat flour, corn flour, steamed rice, steam pretreated willow, starch, and so on. Wheat grain anyway holds the key, and has most normally been utilized, in different procedures.

The choice of a substrate for protein generation in a SSF procedure relies on a few components, predominantly related with expense and accessibility of the substrate, and along these lines may include screening of a few agro-mechanical deposits. In a SSF procedure, the strong substrate not just supplies the supplements to the microbial culture developing in it yet in addition fills in as a safe haven for the cells. The substrate that gives all the required supplements to the microorganisms developing in it ought to be considered as the perfect substrate. In any case, a portion of the supplements might be accessible in problematic focuses, or even missing in the substrates. In such cases, it would wind up important to enhance remotely with these supplements. It has additionally been a training to pretreat (artificially or precisely) a portion of the substrates before their utilization in SSF forms (e.g., ligno-cellulose), in this way making them all the more effectively open for microbial development [20].

Among the few factors that are significant for microbial development and compound generation utilizing a specific substrate, molecule size, starting dampness level, and water movement are the most basic. By and large, littler substrate particles give a bigger surface territory to microbial assault and, in this manner, are an attractive factor. In any case, too little a substrate molecule may bring about substrate agglomeration, which may meddle with microbial breath/air circulation, and accordingly, bring about poor development. Conversely, bigger particles give better breath/air circulation effectiveness (because of expanded interparticle space); yet give restricted surface zone to microbial assault. This requires a trade-off in molecule size for a specific procedure.

Throughout the years, various sorts of fermenters (bioreactors) have been utilized for different purposes in SSF frameworks. SSF procedures are particular from SmF refined, since microbial development and item arrangement happen at or close to the outside of the strong substrate molecule having low dampness substance. Along these lines, it is critical to give advanced water substance and control the water movement (aw) of the maturing substrate, in light of the fact that the benefit capacity of water in lower or higher focuses antagonistically influences microbial action. Also, water profoundly affects the physico-compound properties of the solids and this, thusly, influences the profitability of the general procedure.

5.3.3   STRAIN IMPROVEMENTS

It is very much perceived that the vast majority of naturally occurring micro-organisms don't deliver catalysts at modernly calculable amounts or regularly don't have attractive properties for applications. Subsequently, huge efforts have been made in improving the strains utilizing old-style or sub-atomic instruments to acquire hyper creating strains or building up the required attributes [21–28].

5.3.3.1   MUTATION

The majority of the strains utilized for modern chemical creation have been improved by old-style determination. There are four classes of transformations: (1) spontaneous mutation, (2) error-prone translational synthesis, (3) errors presented during DNA repair, and (4) induced mutation caused by mutagens. Researchers may likewise purposely present freak arrangements through DNA control for logical experimentation. Mutagenesis by UV radiation or compound mutagens has been connected to rapidly locate the valuable variations. Numerous cells are exposed to transformation and the subsequent freaks are chosen for the ideal blend of attributes. More often than not, transformation causes changes of protein structure which results in decay of capacity. Once in a while, changes in basic segments by transformation bring about upgrades except if the particular loss of capacity is required for generation purposes, for instance, when lost administrative capacity brings about improved chemical creation. Transformation and determination are coordinated essentially toward higher by and large profitability as opposed to change of a particular capacity; however, lost administrative capacity is profoundly likely. There are a few instances of freak strains which are referred to as hyper-makers, for example, *Trichoderma reesei* RUT C-30, which has been a standout amongst other cellulase makers for a considerable length of time.

5.3.3.2   RECOMBINANT DNA (RDNA) TECHNOLOGY

A few microorganisms have the ability of delivering the ideal catalyst. Others could win the Olympic gold award in development and catalyst creation. By joining the best from every living being, one could acquire a microorganism that develops in all respects rapidly on reasonable supplements, while simultaneously delivering huge amounts of the correct catalyst. This is finished by recognizing the quality that codes for the ideal chemical and moving it to a creation living being known to be a decent compound maker.

Modern chemicals should be consummately fit to the errands that they perform, however here and there the ideal compound for a particular activity is difficult to discover. This does not mean, in any case, that we can't make a chemical for the activity. Regularly, researchers can discover a normally happening protein that is practically flawless, and utilizing current biotechnology, it tends to be moved up to the ideal effectiveness. This is finished by changing little pieces of the qualities in the microorganism which code for the generation of the chemical. These modest adjustments just change the structure of the chemical all around marginally, yet this is regularly enough to make a decent protein into an ideal catalyst.

Microorganisms segregated from various conditions speak to wellsprings of catalysts that can be utilized for modern procedure science. Despite the fact that the utilization of high-throughput screening (HTS) techniques have empowered us to discover novel and intense proteins from microorganisms, a large number of those microorganisms are not effectively developed in research center conditions or their chemical yield is unreasonably low for practical use. Utilizing DNA innovation, cloning the qualities encoding these proteins and heterologous articulation in usually utilized mechanical strains has turned into a typical practice.

The epic chemicals reasonable for the particular conditions might be acquired by hereditarily changing the microorganism. Modern generation of insulin is delivered by hereditarily adjusted E. coli. Recombinant DNA innovation empowers the generation of chemicals at levels 100-overlay more prominent than the local articulation, making them accessible requiring little to no effort and in enormous amounts. Thus, a few significant nourishment preparing compounds, for example, amylases, and lipases with properties custom fitted to specific sustenance applications have turned out to be accessible. A few microbial strains have been built to expand the catalyst yield by erasing local qualities encoding the extracellular proteases. In addition, certain parasitic creation strains have been changed to diminish or dispose of their potential for the generation of harmful optional metabolites.

Despite the fact that the utilization of DNA innovation altogether brings down the expense of protein creation, the uses of compounds are as yet constrained. Most synthetic concoctions with modern intrigue are not common substrates for these catalysts. On the off chance that an ideal compound action is discovered, the yield is regularly low. Besides, compounds are not generally stable in brutal response conditions, for example, pH higher or lower than physiological pH 7, high temperature, or the nearness of natural solvents required to solubilize numerous substrates. This methodology blocks the exchange of any superfluous or unidentified DNA from the contributor living beings to the generation strain.

5.3.3.3 PROTEIN ENGINEERING

Recent advances in PCR innovation, site-explicit, and irregular mutagenesis are promptly accessible to improve catalyst soundness in a more extensive scope of pH and temperature and resistance to an assortment of natural solvents. Since a huge amount of chemical can be acquired by recombinant articulation, X-beam crystallography can encourage the comprehension of the tertiary structure of a catalyst and its substrate official/acknowledgment locales. This data may help a normal plan of the protein, anticipating amino corrosive changes for adjusting substrate particularity, reactant rate, and enantioselectivity (on account of chiral compound union). To build a monetarily accessible chemical to be a superior mechanical impetus, two unique methodologies are by and by accessible: an arbitrary technique called coordinated development and a protein building strategy called sane structure.

Protein building is a strategy for changing a protein grouping to accomplish an ideal outcome, for example, an adjustment in the substrate particularity, or expanded strength to the temperature, natural solvents, as well as boundaries of pH. Numerous particular strategies for protein building exist, however, they can be gathered into two noteworthy classifications: those including the reasonable structure of the protein changes, and the combinatorial techniques which make changes in a progressively irregular manner.

Protein building or judicious techniques, for example, site-coordinated mutagenesis, require focused on amino corrosive substitutions, and accordingly, require a huge assortment of information on the biocatalyst being improved, including the three-dimensional structure and the synthetic system of the response. The fundamental favorable position of reasonable plan is that few protein variations are made, implying that almost no exertion is important to screen for the improved properties. The combinatorial strategies, then again, make an enormous number of variations that must be tested; in any case, they have the upside of not requiring such broad information about the protein. Furthermore, regularly nonobvious changes in the protein grouping lead to enormous upgrades in their properties, which are amazingly difficult to foresee sanely, and in this way, must be recognized by the combinatorial techniques. A few catalysts have just been built to capacity better in the modern procedures. These incorporate the proteinases, lipases, cellulases, α-amylases, and glucoamylases.

5.3.4 DOWNSTREAM PROCESSING/ENZYME PURIFICATION

The objective of the aging procedures is to create a last detailed protein item and it likewise incorporates many post aging unit tasks. Still, the greatest generation rate could be the most significant factor. Be that as it may, the most minimal unit creation cost could likewise be a significant main impetus. Enhancement of every individual unit activity does not generally prompt an ideal in general process execution, particularly when there are solid communications between unit tasks. A comprehension of these associations is essential to generally speaking procedure advancement. For example, item fixation or immaculateness in the maturation soup can essentially affect downstream purging unit activity. Typically, the point of the refinement procedure is to accomplish most extreme conceivable yield, greatest reactant action, and the greatest conceivable virtue. A large portion of the modern compounds delivered are extracellular and the initial phase in their sanitization is division of cells from the maturation soup. For intracellular proteins, interruption of cells by mechanical or non-mechanical strategies is required. Filtration, centrifugation, flocculation, floatation, lastly focus strategies lead to the advancement of a concentrated item. Salting out and dissolvable precipitation strategies could be utilized for protein focus in businesses. $(CH_3)_2CO$ precipitation is a prevalent technique for protein fixation in ventures as $(CH_3)_2CO$ can be reused. Ultrafiltration, electrophoresis, and chromatography lead to a profoundly decontaminated item.

5.3.5   GENETIC ENGINEERING OF MICROORGANISMS FOR BIOTECHNOLOGY [29–33]

Molecule genetics qualities can be utilized to control qualities so as to adjust the articulation and generation of microbial items, including the declaration of novel recombinant proteins. The exacerbates that are disconnected from plants or creatures can be integrated by hereditary control of various smaller scale life forms to upgrade the generation and by natural and different controls, even up to 1000-crease for little metabolites can be expanded.

The approach of recombinant DNA innovation has significantly widened the range of microbial hereditary controls. With the headway of recombinant DNA innovation, numerous novel host frameworks have been investigated to create economically significant items like remedial proteins, anti-infection agents, little atoms, biosimilars, and so on.

The premise of this innovation is the utilization of limitation endonucleases, polymerases, and DNA ligases as a way to explicitly reorder pieces of DNA. So also, remote DNA pieces can be brought into a vector atom (a plasmid or a bacteriophage), which empowers the DNA to duplicate after presentation into a bacterial cell.

The capacity to change and clone qualities quickened the pace of disclosure and the advancement in biotech businesses. The essential strides in DNA cloning include the accompanying:

• A section of DNA is embedded into a bearer DNA particle, called a vector, to deliver a recombinant DNA.

- The recombinant DNA is then brought into a host cell, where it can duplicate and create various duplicates of itself inside the host. The most usually utilized host is the microscopic organisms, albeit different hosts can likewise be utilized to engender the recombinant DNA.
- Further enhancement of the recombinant DNA is accomplished when the host cell separates, conveying the recombinant DNA in their descendants, where further vector replication can happen.
- After countless divisions and replications, a province or clone of indistinguishable host cell is created, conveying at least one duplicates of the recombinant DNA.
- The state conveying the recombinant DNA of intrigue is then distinguished disengaged, investigated sub-refined and kept up as a recombinant strain.

There are a few techniques by which hereditary modifications of maker microbial strains should be possible for amplification of items or metabolites (Figure 5.1).

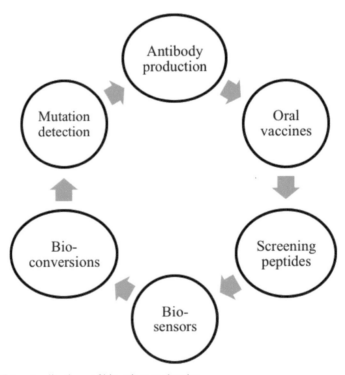

FIGURE 5.1   Applications of bio-micro molecules.

5.3.6   MICROBIAL CELL-SURFACE DISPLAY

Cell-surface display permits proteins, peptides, and other bio full-scale particles to be shown on the outside of microbial cells by melding them with the securing themes. The protein to be display-passenger protein-can is intertwined to tying down theme-the transporter protein-by N-terminal combination, C-terminal combination, or a sandwich combination. The particular highlights of bearer protein, traveler protein, and the host cell, and different combination strategies influence the productivity of surface show-case of bio-macromolecules. Microbial cell-surface showcase has numerous potential applications, including live antibody advancement, peptide library screening, bioconversion utilizing entire cell biocatalyst, and bio-adsorption.

5.3.7   BIOREACTORS

In prior occasions, bioreactors were utilized for a considerable length of time to make wine and lager. The bioreactor is potentially the most significant single bit of hardware utilized in biotechnology. Bioreactors are the compart-ments or vessels that enable organic procedures to happen under ideal conditions. These reactors, in controlled conditions, will yield a valuable substance in enormous sums [34–37].

5.3.8   CELL FUSION

This system includes the combination of two cells to make a solitary cell that contains all the hereditary material of the first cells. Up until now, this procedure has been utilized to make new plants by combining cells from species that don't normally hybridize (from a cross-breed) and afterward producing entire plants from the fused cells.

5.3.9   LIPOSOME-BASED DELIVERY

Liposomes are minute round structures that create when lipids structure and suspension in water. These round vesicles organize themselves in order to create a modest space inside the focal point of the liposome. Such space can conceivably be misused to convey/transport another substance, for example, a medication.

5.3.10   CELL OR TISSUE CULTURE

This method permits the development and division of individual cells in a shower of clean, nutritive liquid which regularly contains hormones and development substances. This technique is utilized widely in organic labs, for instance, in malignancy look into, plant rearing and routine investigation of chromosome karyotypes. The entire procedure is directed in an in vitro condition by giving a reasonable culture medium that contains a blend of supplements either in a strong structure or in fluid-structure [38].

5.3.11   DNA FINGERPRINTING

DNA fingerprinting is a system that is utilized for recognizing the segments of DNA (the material of the qualities) that are special to a specific person. Varieties in DNA among various people can be utilized for recognizable proof purposes. This little area of the DNA of a living being interestingly recognizes that specific life forms from all others. Such differing bits of hereditary material appear as successions of DNA called scaled-down satellites, which are rehashed a few times. The quantity of reiterations of scaled-down satellites per area of quality can shift tremendously between irrelevant people [39].

5.3.12   CLONING

The method of production of identical animals, plants, or smaller scale living beings from a solitary individual is known as cloning. As such, it is a procedure by which a living being is gotten from a solitary parent through non-sexual multiplication. Cloning is skilled in nature to those living beings that repeat agamically and produce their own clones, for example, plants, small scale life forms, and straightforward creatures, for example, corals. Be that as it may, warm-blooded creatures imitate explicitly, and can't clone normally since the relative of a vertebrate acquires its hereditary material not from one parent yet half from each parent. Subsequently, the posterity delivered is never an indistinguishable duplicate of both of its folks. In nature, clones from well-evolved creatures are restricted to the generation of indistinguishable twins [40].

5.3.13   ARTIFICIAL INSEMINATION AND ET TECHNOLOGY

Advancement in the investigation of embryology, urology, and urogyne-cology has prompted progress in the zone managing manual semen injection. Managed impregnation permits the counterfeit presentation of semen into the conceptive tract of a female creature, and is broadly utilized in reproducing creatures, for example, sheep, and dairy cattle. Males with prevailing and alluring inherited attributes/qualities are chosen for semen accumulation. Gathered semen from guys with attractive qualities can be solidified and shipped long separations to treat female creatures. Managed impregnation is additionally utilized to help ladies who wish to consider where ordinary origination is beyond the realm of imagination [41].

5.3.14   STEM CELL INNOVATION

With the headway of biotechnology, it is currently conceivable to use the capability of foundational microorganisms for gainful purposes. Undeveloped cells are undifferentiated and can mitotically partition to make develop utilitarian cells, for example, bone marrow foundational microorganisms can offer ascent to the whole scope of insusceptible framework platelets. Foundational microorganisms are found in many life forms, however, are normally found in multicellular living beings. In 1908, Alexander Maksimov began the term 'undifferentiated cell,' and later foundational microorganism research work was proceeded by Canadian researchers Ernest A. McCulloch and James E. Till during the 1960s. During this period, two broad sorts of mammalian undeveloped cells were found: embryonic undifferentiated organisms that are separated from the internal cell mass of blastocysts, and grown-up immature microorganisms that are found in grown-up tissues. During the improvement of developing lives, foundational microorganisms can separate into the majority of the specific embryonic tissues though, in grown-up creatures, undifferenti-ated cells and ancestor cells go about as a fix framework for the body, renewing particular cells, yet in addition keeping up the ordinary turnover of regenerative organs, for example, blood, skin or intestinal tissues. Momentum research uses exceptionally plastic grown-up undeveloped cells from an assortment of sources, including umbilical line blood and bone marrow, for different medicinal treatments. Progressions in remedial cloning permit the advancement of embryonic cell lines and autologous embryonic foundational microorganisms all the more helpfully and offer the promising possibility for future treatments.

The old-style meaning of an undeveloped cell necessitates that it has two properties:

- Self-reestablishment or the capacity to experience various cycles of cell division while keeping up an undifferentiated state; and
- Potency or the ability to separate into particular cell types [42–44].

5.4 OPPORTUNITIES, CHALLENGES, AND THE IMPORTANCE OF BIOTECHNOLOGY

Biotechnology is the study of the controlled utilization of organic specialists for helpful use. Since biotechnology isn't a free discipline, its outstanding joining with associated fields, for example, natural chemistry, sub-atomic science, and microbiology encourage the mechanical use of organic operators. Along these lines, present-day biotechnology has created as a science with tremendous potential for human welfare in regions going from nourishment preparing to human wellbeing and ecological assurance. The real essentialness of this field of science in various fields will be clear from the accompanying models.

5.4.1 BIOTECHNOLOGY IN MEDICINE

One of the significant regions in biotechnology is the restorative segment. This is the field wherein the greater part of the exploration is occurring and a few achievements have been made. It is likewise the territory that raises the most elevated number of moral and legitimate issues. The extent of biotechnology in medication is to use strategies in living frameworks to deliver remedial proteins, which are generally called biopharmaceuticals or recombinant proteins. Items, for example, monoclonal antibodies, DNA, and RNA tests are delivered for the finding of different illnesses. Moreover, helpful protein-based medications, for example, insulin, and interferon have been combined with microorganisms for the treatment of human ailments. As recently referenced, the utilization of biotechnology in the field of the drug is otherwise called 'red' biotechnology. It manages many major and minor parts of human life, from making drugs increasingly powerful as far as expense and proficiency, to handling one of the most troublesome parts of medication, relieving hereditary ailments. Red biotechnology covers different potential meds for maladies, for example, malignant growth, and AIDS.

The subsequent significant field of red biotechnology is quality treatment, which manages the finding and treatment of hereditary ailments and some different illnesses, for example, malignancy. This treatment incorporates the control of qualities and the rectification of blemished qualities. During this procedure, qualities are embedded, erased, or adjusted. One of the most well-known types of quality treatment is the fuse of useful qualities into an unknown genomic area so as to supplant a changed and useless quality. Pharmacogenomics and hereditary testing both use strategies of red biotechnology that are singular explicit. In pharmacogenomics, the hereditary data of the individual is inferred, and medications are built up that can be embedded into that specific individual, while in hereditary testing various tests are led among relatives to decide hereditary ailments, sex, and transporter screening. It can likewise be utilized in paternity questions. Monoclonal antibodies, DNA, and RNA tests are utilized for the finding of different ailments and profitable medications, for example, insulin, and interferon have been combined by microscopic organisms for the treatment of human illnesses. DNA fingerprinting is used for the distinguishing proof of guardians and offenders.

5.4.2   INDUSTRIAL BIOTECHNOLOGY

Industrial biotechnology was established for the large-scale production of alcohol and antibiotics by micro-organisms. Currently, various pharmaceutical drugs and chemicals such as lactic acid, glycerine, etc., are being produced by genetic engineering for better quality and quantity. Biotechnology has provided us with a very efficient and economical technique for the production of a variety of biochemicals, e.g., immobilized enzymes. Protein engineering is another important area where existing proteins and enzymes are remodeled for a specific function or to increase the efficiency of their function.

5.4.3   BIOTECHNOLOGY AND THE ENVIRONMENT

Ecological issues, for example, contamination control, the exhaustion of regular assets for non-sustainable power sources, protection of biodiversity, and so forth, are being managed utilizing biotechnology. For instance, micro-organisms are being used for the detoxification of mechanical effluents, to battle oil slicks, for treatment of sewage, and for biogas creation. Biopesticides offer earth a more secure option in contrast to compound pesticides for control of creepy-crawly vermin and ailments.

5.4.4   BIOTECHNOLOGY AND AGRICULTURE

Presently the capability of plant tissue culture is broadly used for the quick and financial clonal increase of leafy foods trees, for the generation of virus-free hereditary stock and planting material, just as in the production of novel hereditary varieties through somaclonal variety. With the guide of rDNA innovation, it has now turned out to be conceivable to deliver transgenic plants with alluring qualities, for example, herbicide obstruction, ailment opposition, expanded timeframe of realistic usability, and so on. Strategies, for example, atomic rearing has been utilized to quicken the procedure of yield improvement. For example, sub-atomic markers, for example, confinement section length polymorphism (RFLP: restriction fragment length polymorphism), and straightforward arrangement rehashes (SSRs) give potential devices to the roundabout determination of both subjective and quantitative attributes, and furthermore for contemplating genotypic assorted variety.

5.5   CONCLUSION

Despite the fact that protein innovation is an entrenched part of science, it is as yet going through a constant period of advancement. Our general public is moving towards an eco-accommodating innovation supplanting a few compound based advances in order to ensure our condition for the future age. Looks for novel compounds dependent on a potential application for a realized chemical are pushing forward at the same time. Chemicals have just demonstrated an enormous ability to direct us towards organic procedures as biocatalysts. A few substance procedures have been supplanted by organic procedures with a few advantages, for example, gentle working conditions, explicitness, and ecological achievability. Proteins have made mediations in practically all the real business segments, particularly in the pharmaceutical and nourishment enterprises. Proteins will proceed with their potential and helpful jobs in a progressively increased way later on.

KEYWORDS

- generally recognized as safe
- high-throughput screening
- microbial biotechnology
- micro-organism
- restriction fragment length polymorphism
- solid state fermentation

REFERENCES

1. Bud, R., (1989). Janus-faced biotechnology: An historical perspective. *Trends Biotechnol., 7*, 230–233.
2. Goodman, D. C., (1987). *From Farming to Biotechnology: A Theory of Agro-Industrial Development*. Blackwell Publishers, Oxford.
3. Verma, A. S., Agrahari, S., Rastogi, S., & Singh, A., (2011). Biotechnology in the realm of history. *J. Pharm. Bioallied. Sci., 3*, 321–323.
4. Ereky, K., (1919). Biotechnology of meat, fat, and milk production in the large agricultural holdings: written for scientifically educated farmers (Berlin: Parey).
5. Fári, M. G., & Kralovánszky, U. P., (2006). The founding father of biotechnology: Karl Ereky. *Int. J. Hort. Sci., 12*, 9–12.
6. Campbell, C. S., (2003). Biotechnology and the fear of Frankenstein Camb. *Q. Healthc. Ethics, 12*, 342–352.
7. Belt, H., (2009). Playing god in Frankenstein's footsteps: Synthetic biology and the meaning of life. *Nanoethics, 3*, 257–268.
8. Demain, A. L., (1990). Achievements in microbial technology. *Biotechnol. Adv., 8*, 291–301.
9. Demain, A. L., (1988). Contributions of genetics to the production and discovery of microbial pharmaceuticals. *Pure Appl. Chem., 60*, 833–836.
10. Pandey, A., Webb, C., Soccol, C. R., & Larroche, C., (2006). *Enzyme Technology Springer Science*. USA.
11. Pitman, S., (2011). *Growth in Enzyme Market Driven by Cosmetic Demand*. Online news, cosmetics design. http://www.cosmeticsdesign-asia.com (accessed on 29 June 2020).
12. Rehm, H. J., & Reed, G., (1985). *Biotechnology, a Comprehensive Treatise in EightVolumes*. VCH Verlagsgesellschaft, Weinheim, Germany.
13. Research, & Markets, (2011). In: *Report–Chinese Markets for Enzymes*.
14. Research, & Markets, (2011). In: *Report–Indian Industrial Enzymes Market*.
15. Rich, J. O., Michels, P. C., & Khmelnitsky, Y. L., (2002). Combinatorial biocatalysis. *Curr. Opin. Chem. Biol., 6*, 161–167.
16. Sarrouh, B., Santos, T. M., Miyoshi, A., Dias, R., & Azevedo, V., (2012). Up-to-date insight on industrial enzymes applications and global market. *J. Bioprocess Biotech., S4*, 002.

17. Selvakumar, P., Ashakumary, L., & Pandey, A., (1996). Microbial synthesis of starch saccharifying enzyme in solid-state fermentation. *J. Sci. Ind. Res., 55*(5/6), 443–449.

18. Shu-Jen, C., (2004). Strain improvement for fermentation and biocatalysis processes by genetic engineering technology. *J. Ind. Microbiol. Biotechnol., 31*, 99–108.

19. Singhania, R. R., Patel, A. K., & Pandey, A., (2010). The industrial production of enzymes. In: Soetaert, W., & Vandamme, E. J., (eds.), *Industrial Biotechnology* (pp. 207–226). Wiley-VCH Verlag, Weinheim, Germany.

20. Bibb, M., Ward, J. M., & Hopwood, D. A., (1978). Transformation of plasmid DNA into *Streptomyces* at high frequency. *Nature, 274*, 398–400.

21. Bibb, M. J., & Cohen, S. N., (1982). Gene expression in *Streptomyces*: Construction and application of promoter-probe plasmid vectors in *Streptomyces*lividans. *Mol. Gen. Genet., 187*, 265–277.

22. Brownell, G. H., Saba, J. A., Denniston, K., & Enquist, L., (1981). The development of a Rhodococcus-actinophage gene cloning system. *Dev. Industrial Microbiol., 23*, 287–303.

23. Challis, G. L., (2006). Engineering *Escherichia coli*to produce non-ribosomal peptide antibiotics. *Nature Chemical Biology, 2*, 398–400.

24. Chang, S., & Cohen, S. N., (1979). High frequency transformation of *Bacillus subtilis* protoplasts by plasmid DNA. *Mol. Gen. Genet., 168*(1), 111–115.

25. Ehrlich, S. D., (1977). Replication and expression of plasmids from *Staphylococcus*aureus*in *Bacillus subtilis. Proc. Natl. Acad. Sci., 74*, 1680–1682.

26. Ehrlich, S. D., (1978). DNA cloning in *Bacillus subtilis. Proc. Natl. Acad. Sci., 75*, 1433–1436.

27. Gruenewald, S., Mootz, H. D., Stehmeier, P., & Stachelhaus, T., (2004). *Appl. Environ. Microbiol., 70*, 3282–3291.

28. Gryczan, T. J., Contente, S., & Dubnau, D., (1978). Characterization of *Staphylococcus aureus* plasmids introduced by transformation into *Bacillus subtilis. J. Bacteriol., 134*, 318–329.

29. Hopwood, D. A., Chater, K. F., Dowding, J. E., & Vivian, A., (1973). Advances in *Streptomyces coelicolor* genetics. *Bacteriol. Rev., 37*, 371–401.

30. Katsumata, R., Oka, T., & Furuya, A., (1982). *Novel Lysozyme Sensitive Microorganism.* European Patent Application 92103677.9.

31. Peiru, S., Menzella, H. G., Rodriguez, E., Carney, J., & Gramajo, H., (2005). Production of the potent antibacterial polyketide erythromycin C in *Escherichia coli. Appl. Environ. Microbiol., 71*, 2539–2547.

32. Pfeifer, B. A., Admiraal, S. J., Gramajo, H., Cane, D. E., & Khosla, C., (2001). Biosynthesis of complex polyketides in a metabolically engineered strain of *E. coli. Science, 291*, 1790–1792.

33. Schmidt, E. W., et al., (2005). *Proc. Natl. Acad. Sci. USA, 102*, 7315–7320.

34. Jaeger, K. E., Eggert, T., Eipper, A., & Reetz, M. T., (2001). Directed evolution and the creation of enantioselective biocatalysts. *Appl. Microbiol. Biotechnol., 55*, 519–530.

35. Kuraishi, C., Yamazaki, K., & Susa, Y., (2001). Transglutaminase: Its utilization in the food industry. *Food Rev. Int., 17*, 221–246.

36. Olempska-Beer, Z. S., Merker, R. I., Ditto, M. D., & DiNovi, M. J., (2006). Food-processing enzymes from recombinant microorganisms: A review. *Regul. Toxicol. Pharm., 45*, 144–158.

37. Otten, L. G., & Quax, W. J., (2005). Directed evolution: Selecting today's biocatalysts. *Biomol. Eng., 22*, 1–9.

38. Pandey, A., (1995). Glucoamylase research: An overview. *Starch/Starke, 47*(11), 439–445.

39. Pandey, A., (2003). Solid-state fermentation. *Biochem. Eng. J., 13*(2/3), 81–84.

40. Pandey, A., Binod, P., Ushasree, M. V., & Vidya, J., (2010). Advanced strategies for improving industrial enzymes. *Chem. Ind. Digest, 23*, 74–84.

41. Pandey, A., Larroche, C., Soccol, C. R., & Dussap, C. G., (2008). *Advances in Fermentation Technology.* Asiatech Publishers, Inc., New Delhi.

42. Pandey, A., Nigam, P., Soccol, C. R., Singh, D., Soccol, V. T., & Mohan, R., (2000). Advances in microbial amylases. *Biotechnol. Appl. Biochem., 31*, 135–152.

43. Pandey, A., Selvakumar, P., Soccol, C. R., & Nigam, P., (1999). Solid-state fermentation for the production of industrial enzymes. *Curr. Sci., 77*(1), 149–162.

44. Pandey, A., & Singhania, R. R., (2008). Production and application of industrial enzymes. *Chem. Ind. Digest, 21*, 82–91.

CHAPTER 6

Pharmaceutical Natural and Synthetic Colorants, Pigments, Dyes, and Lakes: Applications, Perspectives, and Regulatory Aspects

DEBARSHI KAR MAHAPATRA[1] and SANJAY KUMAR BHARTI[2]

[1]*Department of Pharmaceutical Chemistry, Dadasaheb Balpande College of Pharmacy, Nagpur – 440037, Maharashtra, India*

[2]*Institute of Pharmaceutical Sciences, Guru Ghasidas Vishwavidyalaya (A Central University), Bilaspur – 440037, Chhattisgarh, India*

6.1 INTRODUCTION

The vibrant field of pharmaceuticals and cosmeceuticals shares a distinct relationship with colors and coloring agents [1]. It has a complex artistic, physiological, symbolic, psychological, and associative role for humans [2]. The modern researches have suggested that product color also influence the therapeutic efficacy. It has been known from the evidence that "warm colors"; the colors in the red area of the color spectrum drastically evoke emotion, produce anger, discomfort, and hostility. In contrast, the "cool colors"; blue, green, purple, etc., have been perceived to calm the mind, produce sadness or other mixed-feelings [3]. The implementation or translation of the net obtained study results into modern medications has its own consequence [4].

The significance of colorants in a number of pharmaceutical dosage forms such as tablets (core or coating), hard or soft gelatin capsule shells, oral liquids, ointments, gels as well as in the cosmetic preparations such as creams, toothpastes, etc., lies in the swift identification of various strength of drug or formulations with physicochemical characteristics, provides

protection to the light-sensitive active materials, helps in product standardization, and abolish possible errors by the producers (at various manufacturing
stages and distribution stages), prescribers such as doctors and pharmacists,
as well as by the consumers (when the products of analogous manifesta-
tion exist in the inventory of diverse manufacturers) [5, 6]. Colors appeal
to masses which gradually enhance the acceptability among the masses
as it reduces intra- and inter-batch variability of the product. In the eyes
of consumers, especially the children, the brightly colored formulations
(tonics, cough mixtures, powders, ointments, etc.), prepared from colors,
lakes, and pigments play dominant for easy palatability compliance due to
their elegance and attractive appearance [7, 8]. Components such as titanium
dioxide, iron oxides, and aluminum lakes have their own importance in the
pharmacy [9]. For providing visual uniformity, some specific industrial
process-oriented examples can be included such as iron-tints in the calamine,
green colorants in soap where natural oils are applied, lactose colored with
caramel in opium, etc. [10].

6.2   COLORANTS

The colorants can be classified as:

1.   Organic dyes;
2.   Lakes;
3.   Inorganic or mineral colors;
4.   Natural colors or vegetable and animal colors.

6.2.1   IDEAL PROPERTIES OF ANY COLORANT

The ideal properties of a colorant include [11, 12]:

- They must be non-toxic;
- They must be free from harmful impurities;
- They have to free from any taste and odor;
- They must be devoid of pharmacological activities;
- They must be readily available and inexpensive;
- They should have a defined chemical formula so that quality control
  assays must be easily established for routine analysis;

- They must have higher stability on storage and remain unaffected by microbes, temperature, humidity, oxidant, reducer, pH change, and light;
- They must be compatible with active pharmaceutical ingredients (APIs) and other components;
- They should not be adsorbed onto the particulate matter/carrier system;
- They must be readily soluble in water and in some cases must express solubility in organic solvents;
- They must possess a high coloring ability.

6.2.2 DYES

Dyes such as Sunset Yellow, Tartrazine, Erythrosine, Patent Blue V, etc., (Figure 6.1) are pure chemical compounds (~99% homogenous) that express coloring attribute or present fixed strength when dissolved in a liquid component (solvent, liquid mixture, propylene glycol, glycerin, etc.), [13]. They have higher coloring attributes than that of the natural pigments, are much economical, and are commercially available in variant shades. The solution of dyes are prepared in glass-lined tanks with proper mixing using stirrer and suitably filtered to remove any undissolved content to obtain a clear solution [14]. The strength of the dye depends on the pure content of dye whereas the particle size, grinding process, formulation type, drying process, etc., influence the quality aspects of the dye [15].

6.2.3 LAKES

Lakes are the aluminum or calcium salts of water-soluble FDA approved dyes that are an insoluble form of common water-soluble dyes and impart color by dispersion [16]. They are available in the following shades: green-blue, orange, pink-red, yellow, orange-red, and royal blue. The blends provide a variety of hues like yellow, orange, brown, red, green, and purple [17]. In general, they are formed by precipitating and absorbing the content over an insoluble substrate such as alumina hydrate, followed by treating with a soluble aluminum salt. Afterward, the content is purified and dried [18]. The particle size largely influences the strength of the lake as with an increase in the surface area, the tinctorial strength enhances gradually. Furthermore, the preparation technique, manufacturing conditions, dispersibility, concentration, reactant, pH, agitation speed, temperature, etc., have influenced over the final formulation [19]. In comparison to the pure dye content, the shade of the lake varies considerably. Some classic examples include indigo carmine lake, sunset yellow lake, brilliant blue lake, amaranth lake, quinoline yellow lake, and Allura red lake [20]. The applications of aluminum lakes have numerous advantages:

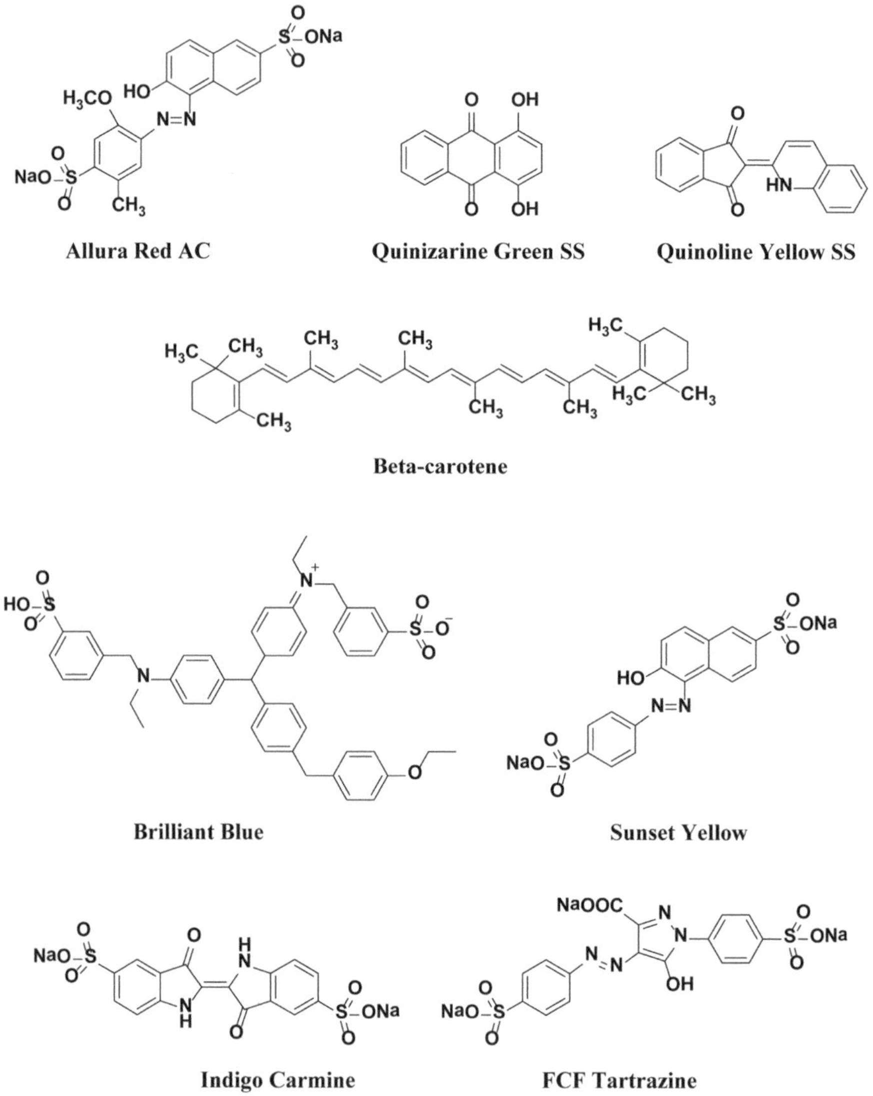

Allura Red AC   **Quinizarine Green SS**   **Quinoline Yellow SS**

Beta-carotene

Brilliant Blue   **Sunset Yellow**

Indigo Carmine   **FCF Tartrazine**

FIGURE 6.1   Structures of some pharmaceutically privileged colorants.

- Full color can be developed by a fewer application of the content which eventually saves time.
- The drying protocols are quicker due to their insoluble nature.
- Over coloring and mottling does not result due to the opacity of the system.
- The cost of raw material improves significantly along with the elimination of several types of defects.

6.2.4   INORGANIC COLORS OR MINERAL COLORS

It involves materials that protect the active ingredients from light and thereby improving their shelf-life. The enhancement of stability of ingredients in a hard gelatin capsule by titanium dioxide is a classic example [21]. The wide regulatory acceptance of inorganic colors provide opportunities to several multinational organizations, however, limited shades (hues) are a major drawback [22]. Until coal tar dyes came into existence, the mineral pigments serve as the primary source for coloring food material, but are now of less significance due to the dominance of synthetic dyes which are less toxic. Currently, it is employed in calamine to impart flesh color (a mixture of yellow and red ferric oxides) [23].

6.2.5   NATURAL COLORS OR VEGETABLE AND ANIMAL COLORS

These are the group of chemically dissimilar component that is either procured from nature through various extractive techniques or synthesized chemically in the laboratory through modified procedures [24]. In the old era, vegetable extracts are regularly employed as diluted preparation. For example, β-carotene which is a very common natural coloring ingredient available from Mother Nature, however, the synthetic origin or often termed as 'nature identical' is widely available due to its low-cost [25]. Even some of the products which are not a part of normal human diets such as annatto or cochineal are also termed as 'natural' which is not at all good [26]. Their wide acceptability did not help them in achieving the hall of fame as they are not stable towards the light. Three ingredients: carmine (aluminum lake of cochineal), cochineal (dried insect), and caramel (black viscid mass of water-soluble carbohydrates) are classic examples of natural coloring agents [27]. Curcumin, riboflavin, annatto, paprika oleoresin, anthocyanins, beetroot red, etc., are other examples.

The main drawbacks of natural vegetable-based colors:

- A variation in coloring attribute is often perceived;
- They are more expensive than coal tar dyes;
- The tinctorial power is very low.

6.3   APPLICATIONS OF DYES, PIGMENTS, AND LAKES IN PHARMACEUTICAL FORMULATIONS

6.3.1   TABLET

6.3.1.1   WET GRANULATION

For coloring any tablet, water-soluble dyes are initially dissolved in the binding solution during the process of granulation [28]. During the manufacturing process, water-soluble dyes migrate and lead to the uneven color of the product. The migration of colors (more than one) at varied rate results in a defect called "mottling" where the distribution of the color is not even in the product and are largely unaccepted by the users [29]. For overcoming such problems, insoluble aluminum lakes or pigments are employed as they do not migrate owing to their insoluble characteristics [30]. The incorporation of pigments and lakes drastically improves the light stability of the formulation. In other ways, water-soluble dyes are adsorbed on clay, starch, talc, etc., for reducing the migration of colors [31].

6.3.1.2   DIRECT COMPRESSION

The direct compression technique is an emerging formulation method that is growing popular due to the elimination of wet steps which considerably enhance the stability of the product [32]. Likewise, in the previous case where the water-soluble dyes have solubility issues and several other considerations, here, pigments, and lakes can be blended directly with other ingredients without any efforts [33]. However, like any industrial process, care has to be taken regarding the color migration, formation of "hot spots" due to poor blending, and color specking [34]. If the problem occurs, it can be eliminated effectually by pre-blending the other ingredients with the pigment before the final addition to the mixture [35]. This step efficiently prevents the pigment particle agglomeration. Though, the ease of incorporation of pigment(s) in the formulation largely varies with the mixture composition [36].

6.3.2   TABLET COATING

6.3.2.1   SUGAR COATING

The coloring of any finished product is a crucial, delicate, and time-consuming operation in the formulation process as it enhances the elegance, provides a definite shape, as well as prevents any visual defect [37]. Several decades back, soluble dyes were employed as the primary colorant for the traditional color coating. Although, the process produces elegant tablets but dye-related difficulties such as migration of color, non-uniform distribution, over coloring, etc., are frequently encountered due to drying time, handling technique, non-optimization, and fabricated topology [38]. For producing a uniform, elegant, brightly-colored product, the surface is initially made smooth by multiple coatings and high concentrations of colored syrup are utilized with multiple applications (20 to 60 times) for preventing mottling [39]. In modern days, pigments, and aluminum lakes are employed in a syrup solution for the coating process which helped in overcoming the anomalies associated with olden-day operations [40]. As the simple rule of nature, every constituent has some merits as well as demerits. The modern process of coating needs air filtration facilities, air conditioning, and dust collecting system to avoid cross-contamination due to the dusty particulate nature of the pigments [41]. With the advance in quality assurance, several companies are producing pre-dispersed, color-matched, and pigment-based sugar-coating concentrates that can be easily dispersed in the sugar solution and can be used for coating tablets in a reproducible manner [42].

6.3.2.2   FILM COATING

The utilization of the film coating overcomes several problems associated with the sugar coating. The film-forming polymers form a thin film on the surface of the tablet, capsule, granule, etc. [43]. For achieving the desired properties of the solid dosage forms; colorants, plasticizers, etc., are often employed which certainly enhances the quality attributes of the product [44]. During the coating operation, the specific polymer and the compatible plasticizer are dissolved in a definite organic solvent containing the colorant

[45]. Due to regulatory restrictions, the organic solvents are replaced to some extent and aqueous systems are applied for coating, however, it is limited to water-soluble dyes only [46]. Due to small thin polymeric film over the tablets, a minute disparity in the thickness of the film will lead to a considerable variation in the color [47]. These pacified systems, however, suffer from poor photostability as compared to the lakes and pigments because the latter reduces the diffusion of moisture through the film [48]. The total film-coating in coating pan has been much easy than before due to the availability of the commercially available pre-dispersed, concentrated, dried forms of coloring materials which can be directly dispersed in aqua [49].

6.3.3 CAPSULE

6.3.3.1 HARD GELATIN CAPSULE

The hard gelatin capsules are primarily colored with FD&C approved water-soluble dyes and colorants and also with opacifier titanium dioxide [50]. During the manufacturing process, the opacifier and coloring materials (dissolved or dispersed in water, glycerin, or any vehicle) are added to the molten gelatin mass, maintaining a proper pH, wall thickness, and tackiness to sustain and maintain the shades of the color in the walls of the product [51]. The colored imprinting edible inks enable the manufacturers to print logo, name, and bands of variable size and design on the body of the product via spin printing technology [52].

6.3.3.2 SOFT GELATIN CAPSULE

The soft gelatin capsules are single-piece sealed shells containing a liquid, semi-solid, suspension, or any constituent of desired state bearing pharmacologically active ingredient [53]. These hermetically sealed shells may contain titanium dioxide as the opacifier (0.5% concentration) for preventing the degradation of the light-sensitive drug as well the shell is usually made darker in color than the color of the encapsulated material [54].

6.3.4 LIQUIDS

In liquids (solution, emulsion, suspension, etc.), the most important criteria for the dyes are that it must be completely soluble in the vehicle in the required concentration to form an opaque liquid without getting precipitated [55]. While formulating any dye-based product, minimum concentration (<0.001%) must be preferred because a large concentration leads to a dull color. As well, a small quantity of undissolved dye material formed during the compounding procedure in the bulk mixing tank may cause trouble in getting a homogenous solution [56]. Sometimes, the formulator chooses colors complementary to the flavor, such as yellow for lemon, orange for orange, red for cherry, etc. Factors such as pH, compatibility, exposure to diffused light, microbiological stability, etc., must be taken care of which directly influences the shade and stability of the colors [57]. During the formulation development of non-aqueous systems, pigments are used charily due to solubility restrictions. Pigments are cautiously pre-dispersed initially before addition to the final product [58].

6.3.5   OINTMENTS

Depending on the base or vehicle, for coloring the ointment formulations, pigments, and dyes are employed. However, the former is most preferred as they do not migrate to the surface which is a common problem with the dyes. The application of the roll or ointment mill leads to the uniform blending of the pigment in the product [59].

6.3.6   TOOTHPASTE

In any toothpaste formulation, the organoleptic appearance of color is a major challenge. In the majority of the cases, migration of color represents a great problem. The problem becomes more serious when a colored component is applied to the white base, therefore, a colorant that can hide the bleeding must be selected judiciously [60]. The problem was duly overcome by high-density polyethylene components that entrain the colorants in both powdered and gel-based toothpaste [61]. In contrast to the low-density polyethylene, polymeric resin, and paraffin wax containing formulations for a similar purpose, the high-density materials serve as a better element for color encapsulation [62].

6.4   REGULATORY GUIDELINES

The legislation of any nation governs the purity and safety aspects of the coloring agents. In regions of the world, the regulatory agencies provide sole status as excipients in pharmaceutical preparations which help in a distinction between food use and medicinal applications. There are several guidelines put forward by the international agencies for complete governing of the colorants, dyes, pigments, and lakes [63].

6.4.1   EUROPEAN UNION LEGISLATION

The Council Directive 78/25/EEC, 1977 of EU is the legislation that principally governs the coloring materials subjected to cosmetics, foods, and pharmaceutical products. At present, Council Directive 94/36/EC is responsible for providing the guidance, safety, utilization on food color (amaranth, canthaxanthin, silver, erythrosine, and aluminum) legislation by providing the cross-references [64]. The Scientific Committee on Medicinal Products and Medical Devices have put forward certain guidelines for the suitability and safety of colorants [65].

6.4.2   UNITED STATES LEGISLATION

The administration of pharmaceutical colorants in food is governed by the Color Additive Amendment to the Food Drug and Cosmetic Act, 1960 in the United States. In addition to it, the Food Drug and Cosmetic, Drug, and Cosmetic, and External Drug and Cosmetic play an imperative role in certifying the colors from the natural origin [66].

6.4.3   LICENSING AUTHORITY APPROVAL

The pharmaceutical licensing authority may impose national and international level restrictions during the approval stages. Colors like tartrazine and azo-based colorants have toxic potentials and may show allergic conditions after chronic administration. The authorities restrict their regular, high-amount usage in the normal course [67].

6.4.4   THE FOOD, DRUG, AND COSMETIC ACT

The Indian Food Drug and Cosmetic Act, 1938 has imposed regulatory guidelines for coal tar dyes, where out of three categories, the first two are of application to chewable tablet manufacturing [68]. In addition to it, the Act provides certifiability of colorants, dyes, and pigments, in foods, cosmetics, and medicinal products which are in contact with the mucous membranes. The Act also prevents the internal usage of colorants due to toxic nature and the product may be applied externally [69].

6.5   FACTORS AFFECTING PRODUCT QUALITY

6.5.1   BLENDING OF COLORANTS

The blending of the colors is the process where individual colorant is blended to produce various shade formation. The process leads to satisfactory hues or shades which attract the consumers and this glorious effect cannot be achieved by a single colorant [70]. The classical example includes when Green S is added to distilled water, a greenish-blue coloration is produced, but as tartrazine is added to the above solution, a more satisfactory green color result [71]. Similarly, brilliant blue FCF produces numerous green shades [72]. The US National Formulary has put forward certain guidelines regarding the proportions of colorants (water and oil soluble) to produce various approved hues for a number of drug preparations [73]. The quality control of the colored formulations is done by spectral studies and physico-chemical characterization [74].

6.5.2   STABILITY AND STORAGE CONDITIONS

Various pharmaceutical products commercially available in the market have varied physicochemical stability. The low-toxicity profile of the colorants plays a key role in their presence in medicinally active formulations [75]. The colorants of various chemical groups have poor (organic colorants) to high stability (inorganic pigments, inorganic dyes, aluminum lakes, and synthetic dyes) under standard conditions such as well-closed condition, ideal humidity, light-resistance, and 30°C temperature [76]. The colors obtained from a natural source or identical sources require more rigorous and the manufacturer recommends specific guidelines for improving the shelf-life of the product [77]. When the conditions are quite stringent, they are frequently encapsulated in gelatin shells and sealed under nitrogen atmosphere. To recompense for the possible losses that happened during the manufacturing process and the color shade started to fade, a slight surplus amount of the product is added at the commencing step, keeping precaution that unattractive hues are not produced [78]. During processing and storage, the following care must be taken from encountering the colorants. It must be protected from the oxidizing agent (hypochlorite, chlorine), a reducing agent (ascorbic acid, sugars, metallic ions), microorganisms (mold, bacteria, fungi), extreme pH, high temperature, humidity, light exposure, etc. [79]. The Code of Federal Regulations 21 CFR includes levels, restrictions, additives (certified and uncertified), etc., which acts as a blueprint for present pharmaceutical industries [80].

6.5.3   SAFETY

International bodies like FDA, WHO, EC, etc., have performed long-term toxicological studies and framed certain guidelines related to the use of colorants in food and drugs after continuous monitoring [81]. Based on the observed adverse toxicological effects, a list of permitted colorants has been issued by the regulatory organizations that can be used widely in food and medicines [82]. Although, these colorants are free from any serious events when used in a small amount but a limited population is associated with adverse effects on larger consumption. The Medical Council has put forward several evidences of hypersensitivity and hyperkinetic activity in children after regular consumption of colorants (Amaranth, Tartrazine, Ponceau 4R, Carmoisine, Brilliant black BN, Sunset yellow) for long duration in foods [83]. Allura Red AC is not recommended for children and is presently banned in France, Sweden, Denmark, Belgium, and Switzerland [84]. Erythrosine Lake is found to induce cancer and is presently banned in the USA for any human or animal use. Although, it is used in food in a small quantity since it does not produce any immediate health hazard [85]. Table 6.1 depicts the status of various colorants in the modern scenario. Tartrazine is found to exhibit hypersensitivity reactions and shown to cause hives in a limited population. Due to controversy over its safety issues, it is currently banned in Norway [86]. Nowaday, researchers are concerned about the safety of absorbable dyes and have developed non-absorbable dyes which has limited absorption characteristic from the human gastrointestinal (GI) tract [87].

6.5.4   HANDLING PRECAUTIONS

The international bodies have directed the color manufacturing industries for safe handling of color products by the workers. While handling the colors,

TABLE 6.1 Regulatory Aspects of Some Pharmaceutically Important Dyes

Color	Color Index No.	Type	Appearance	Uses	Contraindication	Regulatory Status
Allura Red AC	CI 16035	Synthetic	Dark red powder	Use in cosmetics, drugs, and food	Slightly less allergy or intolerance reaction by aspirin intolerant people and asthmatics	Banned in Denmark, Belgium, France, Germany, Sweden, Switzerland, Austria, and Norway
Beta-carotene	CI 75130 (natural) and CI 40800 (synthetic)	Both natural and synthetic	Red crystals	Color for sugar-coated tablets	NA	No ban information
Brilliant Blue FCF	CI 42090	Synthetic	Various shades of green	Used in soaps, shampoos, mouthwashes	Induce allergic reaction in individuals with pre-existing moderate asthma	Banned in Austria, Belgium, Denmark, France, Germany, Greece, Italy, Spain, Norway, Sweden, and Switzerland
Indigo Carmine	CI 73015	Synthetic	Dark blue powder	Manufacturing of capsules, topical preparations	NA	Banned in Norway
Iron Oxides	Iron oxide black (CI 77499) and Iron(III) oxide hydrated (CI 77492)	Both natural and synthetic	Yellow, red, black, or brown powder	Cosmetics, foods, and pharmaceutical applications s colorants and UV absorbers	Kidney damage, blindness, a suspected neurotoxin	Banned in Philippines, Germany
Quinizarine Green SS	CI 61565	Synthetic	Green dye	Cosmetics and medications	NA	NA

TABLE 6.1 (*Continued*)

Color	Color Index No.	Type	Appearance	Uses	Contraindication	Regulatory Status
Quinoline Yellow SS	CI 47000	Synthetic	Bright yellow dye with green shade	Used in spirit lacquers, polystyrene, polycarbonates, polyamides, acrylic resins, and to color hydrocarbon solvents	NA	Banned in Australia, Japan, Norway, and the United States
Sunset Yellow FCF	CI 15985	Synthetic	Reddish yellow powder	Brown coloring in both chocolates and caramel	NA	Banned in Norway and Finland
Tartrazine	CI 19140	Synthetic	Yellow or orange-yellow powder	Aqueous solutions	May cause allergic-type reactions (including bronchial asthma) in certain susceptible persons	Banned in Norway, Austria, and Germany
Titanium dioxide	CI 77891	Synthetic	Silver powder	White pigment and opacifier	NA	Banned in France

NA – not available.

minor precautions are suggested due to their low hazardous nature, however, a large quantity of dusted materials such as aluminum lakes, natural dyes, organic dyes, etc., needs special precautions as inhalation produces occupational diseases [88]. It is made mandatory to make use of mask, slipper, glove, cap, and ventilation while handling [89].

6.6   CONCLUSION

The colorants are the heart of pharmaceutical formulations (tablet, pellet, capsule, liquids, etc.) and food products. The selection of particular coloring agents, utilization in formulations, and their psychological effect of the colors have been witnessed by consumers, scientists, formulators, and manufacturers. The exact selection is a tedious process and requires approval. The long-term toxicity of the colorants, lakes, dyes, and pigments has been comprehensively studied by government agencies and at present, regulatory agencies have put guidelines for the safety issues of the constituents. Several nations have either restricted the usage in food or have completely banned the coloring agents with only a handful in usage. In this modern avenue, the colorants are needed to be judged for their benefit ratio, regulatory restrictions, innovations, discoveries, and safety.

KEYWORDS

- **active pharmaceutical ingredients**
- **colorant**
- **dye**
- **pharmaceutical**
- **physicochemical stability**
- **pigment**

REFERENCES

1. Nassau, K., (1997). *Color for Science, Art and Technology* (Vol. 1, 1st edn.). Elsevier Science.
2. Zollinger, H., (2003). *Color Chemistry: Syntheses, Properties, and Applications of Organic Dyes and Pigments* (3rd edn.). John Wiley & Sons.

3. Whitfield, T. W., & Whiltshire, T. J., (1990). Color psychology: A critical review. *Genetic, Social, and General Psychology Monographs, 116*(4), 385–411.

4. Shahid, M., & Faqeer, M., (2013). Recent advancements in natural dye applications: A review. *Journal of Cleaner Production, 53*, 310–331.

5. Swarbrick, J., (2013). *Encyclopedia of Pharmaceutical Technology* (4th edn.). CRC Press.

6. Rowe, R. C., Paul, S., & Marian, Q., (2009). *Handbook of Pharmaceutical Excipients* (6th edn.). Libros Digitales-Pharmaceutical Press.

7. Collett, D. M., Michael, E. A., & John, W. C., (1990). *Pharmaceutical Practice* (8th edn.). Churchill Livingstone.

8. Lechner, A., Jeffrey, S. S., & Leslie, H., (2012). Color-emotion associations in the pharmaceutical industry: Understanding universal and local themes. *Color Research and Application, 37*(1), 59–71.

9. Rowe, R. C., (1984). The opacity of tablet film coatings. *Journal of Pharmacy and Pharmacology, 36*(9), 569–572.

10. Jones, D. S., (2016). *FAST track Pharmaceutics Dosage Form and Design* (2nd edn.) Pharmaceutical Press.

11. Peters, A. T., & Harold, S. F., (1995). *Modern Colorants: Synthesis and Structure* (Vol. 3, 1st edn.). Blackie Academic & Professional.

12. Sung-Hoon, K., (2006). *Functional Dyes* (1st edn.). Elsevier Science.

13. Pollock, I., Young, E., Stoneham, M., et al., (1989). Survey of colorings and preservatives in drugs. *BMJ, 299*(6700), 649–651.

14. McLaren, K., (1986). *The color Science of Dyes and Pigments* (1st edn.). John Wiley & Sons.

15. Hunger, K., (2007). *Industrial Dyes: Chemistry, Properties, Applications* (4th edn.). John Wiley & Sons.

16. Shobha, R., (2000). *Text Book of Pharmaceutics and Pharmacokinetics* (1st edn.). Prism Publications.

17. Bharath, S., (2013). *Pharmaceutics: Formulations and Dispensing Pharmacy* (1st edn.). Pearson Education India.

18. Downham, A., & Paul, C., (2000). Coloring our foods in the last and next millennium. *International Journal of Food Science and Technology, 35*(1), 5–22.

19. Waring, D. R., & Geoffrey, H., (2013). *The Chemistry and Application of Dyes* (1st edn.). Springer Science & Business Media.

20. Mahmud-Ali, A., Fitz-Binder, C., & Thomas, B., (2012). Aluminum-based dye lakes from plant extracts for textile coloration. *Dyes and Pigments, 94*(3), 533–540.

21. Christie, R., (2014). *Color Chemistry* (2nd edn.). Royal Society of Chemistry.

22. Gordon, P. F., & Peter, G., (2012). *Organic Chemistry in Color* (1st edn.). Springer Science & Business Media.

23. Herbst, W., & Klaus, H., (2006). *Industrial Organic Pigments: Production, Properties, Applications* (3rd edn.) John Wiley & Sons.

24. Aberoumand, A., (2011). A review article on edible pigments properties and sources as natural biocolorants in foodstuff and food industry. *World J. Dairy Food Sci., 6*(1), 71–78.

25. Carmen, S., (2007). New technologies to synthesize. Extract and encapsulate natural food colorants. *Bulletin USAMV–CN, 63*, 64.

26. Sigurdson, G. T., Peipei, T., & Monica, G. M., (2017). Natural colorants: Food colorants from natural sources. *Annual Review of Food Science and Technology, 8*, 261–280.

27. Barnett, J. R., Sarah, M., & Emma, P., (2006). Color and art: A brief history of pigments. *Optics and Laser Technology, 38*(4–6), 445–453.
28. Subramanyam, C. V. S., (2017). *Textbook of Pharmaceutics* (2nd edn.). Vallabh Prakashan.
29. Venkataraman, K., (1971). *The Chemistry of Synthetic Dyes* (Vol. 4, 1st edn.). Elsevier Science.
30. Griffiths, J., (1976). *Color and Constitution of Organic Molecules* (1st edn.). Academic Press.
31. Augsburger, L. L., & Stephen, W. H., (2016). *Pharmaceutical Dosage Forms-Tablets* (3rd edn.). CRC Press.
32. Marriott, J. F., (2010). *Pharmaceutical Compounding and Dispensing* (2nd edn.). Pharmaceutical Press.
33. Wade, A., (1980). *Pharmaceutical Handbook: Incorporating the Pharmaceutical Pocket Book* (1st edn.). Rittenhouse Book Distributors.
34. Carter, S. J., (2008). *Cooper and Gunn's Dispensing for Pharmaceutical Students* (12th edn.). CBS Publishers & Distributors Ltd.
35. Langley, C. A., & Dawn, B., (2012). *Pharmaceutical Compounding and Dispensing*. Pharmaceutical Press.
36. Vijaya, R. C., (1995). *Practical Hand Book of Physical Pharmaceutics* (1st edn.). New Century Book House (P) Ltd.
37. Banker, G. S., Juergen, S., & Christopher, R., (2002). *Modern Pharmaceutics* (1st edn.). CRC Press.
38. Shotton, E., & Kenneth, R., (1974). *Physical Pharmaceutics* (1st edn.). Oxford Clarendon Press.
39. Cherng-Ju, K., (2004). *Advanced Pharmaceutics: Physicochemical Principles* (1st edn.). CRC Press.
40. Denton, P., & Chris, R., (2013). *Pharmaceutics: The Science of Medicine Design* (1st edn.). Oxford University Press.
41. Reife, A., & Harold, S. F., (2000). Pollution prevention in the production of dyes and pigments. *Textile Chemist and Colorist and American Dyestuff Reporter, 32*(1), 56.
42. Waring, D. R., & Geoffrey, H., (2013). *The Chemistry and Application of Dyes* (1st edn.). Springer Science & Business Media.
43. Allen, L., & Howard, C. A., (2013). *Ansel's Pharmaceutical Dosage forms and Drug Delivery Systems*. Lippincott Williams & Wilkins.
44. Perrie, Y., & Thomas, R., (2012). *FAST Track Pharmaceutics: Drug Delivery and Targeting*. Pharmaceutical Press.
45. Bentley, A. O., Walter, A. B., & Ellis, M., (1937). *Text-Book of Pharmaceutics*.
46. Felton, L. A., (2016). *Aqueous Polymeric Coatings for Pharmaceutical Dosage Forms* (4th edn.). CRC Press.
47. Parrott, E. L., (1970). *Pharmaceutical Technology: Fundamental Pharmaceutics*. Burgess Publishing Company.
48. Florence, A. T., & Jürgen, S., (2009). *Modern Pharmaceutics Volume 1: Basic Principles and Systems* (5th edn.). CRC Press.
49. Bouwman-Boer, Y., Fenton-May, V., & Le Brun, P. P. H., (2015). Practical pharmaceutics. *An International Guideline for the Preparation, Care, and Use of Medicinal Products*. Cham [u.a.]: Springer.
50. Kasture, P. V. M., (2008). *Pharmaceutics-I* (1st edn.). Pragati Books Pvt. Ltd.

51. Washington, N., Clive, W., & Clive, W., (2000). *Physiological Pharmaceutics: Barriers to Drug Absorption* (1st edn.). CRC Press.

52. Dash, A., Somnath, S., & Justin, M. T., (2013). *Pharmaceutics: Basic Principles and Application to Pharmacy Practice* (1st edn.). Academic Press.

53. Hadkar, U. B. M., (2008). *Practical Physical Pharmacy and Physical Pharmaceutics* (9th edn.). Pragati Books Pvt. Ltd.

54. Ghosh, T. K., & Bhaskara, M. R. J., (2004). *Theory and Practice of Contemporary Pharmaceutics* (1st edn.). CRC Press.

55. Sinko, P. J., & Yashveer, S., (2006). *Martin's Physical Pharmacy and Pharmaceutical Sciences* (6th edn.). Lippincott Williams & Wilkins.

56. Chengaiah, B., Rao, K. M., & Kumar, K. M., (2010). Medicinal importance of natural dyes a review. *International Journal of Pharm. Tech. Research, 2*(1), 144–154.

57. Felton, L. A., (2013). *Remington-Essentials of Pharmaceutics* (1st edn.). Pharmaceutical Press.

58. Augustijns, P., & Marcus, E. B., (2007). *Solvent Systems and Their Selection in Pharmaceutics and Biopharmaceutics* (1st edn.). Springer Science & Business Media.

59. Davis, H., (1950). *Bentley's Textbook of Pharmaceutics* (1st edn.). *Academic Medicine.*

60. Wong, M., & Michael, P., (1999). *Striped Toothpaste Stable to Color Bleeding.* U.S. Patent No. 5,876,701.

61. Allam, K. V., & Gannu, P. K., (2011). Colorants-the cosmetics for the pharmaceutical dosage forms. *Int. J. Pharm. Pharm. Sci., 3*(3), 13–21.

62. Gaud, R. S., (2008). *A Textbook of Pharmaceutics* (1st edn.). Nirali Prakashan.

63. Mantus, D. S., (2014). The practice of regulatory affairs. *FDA Regulatory Affairs* (3rd edn.). CRC Press.

64. Faulkner, E. B., & Russell, J. S., (2009). *High Performance Pigments* (1st edn.). John Wiley & Sons.

65. Šuleková, M., Smrčová, M., Hudák, A., et al., (2017). Organic coloring agents in the pharmaceutical industry. *Folia Veterinaria, 61*(3), 32–46.

66. Marmion, D. M., (1991). *Handbook of US Colorants: Foods, Drugs, Cosmetics, and Medical Devices* (3rd edn.). John Wiley & Sons.

67. Al-Achi, A., Mali, R. G., & William, C. S., (2013). *Integrated Pharmaceutics: Applied Preformulation, Product Design, and Regulatory Science* (1st edn.). John Wiley & Sons.

68. Fleischer, M., Kelm, S., Palm, D., et al., (2000). *Regulation and Innovation in the Chemical Industry*. Joint Research Centre, European Commission.

69. Gregory, P., (2000). Dyes and dye intermediates. *Kirk-Othmer Encyclopedia of Chemical Technology.*

70. Waring, D. R., & Geoffrey, H., (1990). *The Chemistry and Application of Dyes* (1st edn.). Springer.

71. Luo, R., (2016). *Encyclopedia of Color Science and Technology* (1st edn.). Springer Science and Business Media.

72. Mahapatra, N. N., (2016). *Textile Dyes* (1st edn.). CRC Press.

73. Reife, A., (2000). Dyes, environmental chemistry. *Kirk-Othmer Encyclopedia of Chemical Technology.*

74. Best, J., (2012). *Color Design* (1st edn.). Woodhead Publishing.

75. Shindy, H., (2016). Basics in colors, dyes, and pigments chemistry: A review. *Chem. Int., 2*(1), 41–47.

76. Gürses, A., Açıkyıldız, M., & Güneş, K., (2016). *Dyes and Pigments* (1st edn.). Springer.

77. Anantharaman, A., Subramanian, B., Chandrasekaran, R., et al., (2014). Colorants and cancer: A review. *Industrial Crops and Products, 53*, 167–186.

78. Venkatraman, K., (1978). *The Chemistry of Synthetic Dyes* (Vol. 8, 1st edn.). Academic Press.

79. Reife, A., Abraham, R., & Harold, S. F., (1996). *Environmental Chemistry of Dyes and Pigments* (1st edn.). John Wiley & Sons.

80. Clarke, E. A., & Steinle, D., (1995). Health and environmental safety aspects of organic colorants. *Review of Progress in Coloration and Related Topics, 25*, 1–5.

81. Golka, K., Silke, K., & Zdislaw, W. M., (2004). Carcinogenicity of azo colorants: Influence of solubility and bioavailability. *Toxicology Letters, 151*(1), 203–210.

82. Rymbai, H., Sharma, R. R., & Manish, S., (2011). Bio-Colorants and its implications in health and food industry: A review. *International J. of Pharm. Tech. Research, 3*(4), 2228–2244.

83. Chafee, F. H., & Guy, A. S., (1967). Asthma caused by FD& C approved dyes. *Journal of Allergy, 40*(2), 65–72.

84. Florence, A. T., (2010). *An Introduction to Clinical Pharmaceutics* (1st edn.). Pharmaceutical Press.

85. Ibraheem, N. A., Hasan, M. M., Khan, R. Z., et al., (2012). Understanding color models: A review. *ARPN Journal of Science and Technology, 2*(3), 265–275.

86. Bernstein, I. L., Gallagher, J. S., Johnson, H., et al., (1980). Immunologic and nonimmunologic factors in adverse reactions to tartrazine. *Proceedings, 4th FDA Science Symposium* (pp. 258–260).US Govt. Printing Office, Washington, DC.

87. Bechtold, T., & Rita, M., (2009). *Handbook of Natural Colorants* (1st edn.). John Wiley & Sons.

88. Heaton, C. A., (1994). *The Chemical Industry* (1st edn.). Springer, Dordrecht.

89. Bergfeld, W. F., Belsito, D. V., Marks, J. G., et al., (2005). Safety of ingredients used in cosmetics. *Journal of the American Academy of Dermatology, 52*(1), 125–132.

CHAPTER 7

Microbial Pigments: A Green Microbial Technology

EKNATH AHIRE,[1] SWATI G. TALELE,[2] and GOKUL S. TALELE[3]

[1]Divine College of Pharmacy, Nanpur Road, Satana, Nasik, Maharashtra, India

[2]Department of Pharmaceutics, Sandip Institute of Pharmaceutical Sciences, Mahiravani, Nashik, Maharashtra, India

[3]Matoshree College of Pharmacy, Eklahare, Nashik, Maharashtra, India

7.1 ORIGIN OF MICROBIAL PIGMENTS

Color in foods as well as in pharmaceutical products contributes a significant role to shown products freshness and improved external appearance and increased acceptance towards the consumers. The uses of the natural origin of such compounds are growing vastly due to the consciousness of health benefits from natural compounds. Hence, it is required to discover the different natural origins of food class colorants and their capabilities. Among so many available natural origin colorants, microbial pigments, shows an important role as coloring constituents, due to its easiness in production. At industrial level manufacture of the natural colorants from microbe's fermentation process, have lots of advantages like cost-effective production, increased product yield, ease to extraction, and no shortage of raw material and no dependency on seasonal material variants [1]. Indeed, colors are got from the color-producing compounds observe in the living system, called pigment. Pigments multiplicity uses based on modifications in their presence of particular chromophore and chemical structures. After a long period, we obtained pigment from natural bases and their usage has demanded because of the disadvantages produced with the toxic nature of the synthetic colors. In this sense, the microbial pigments are a worthy alternative. Microbes are the most significant living being in survival and govern the lifecycle and demise on this globe. Microorganism is mostly accompanying with all types of eatable food and also they are accountable for the formation of various food products. They also provide the good quality of pigments, vitamins, amino acids, different enzymes, and organic acids, etc. [2]. Pigments are the substances by characteristics of significant for many producer industries. Pigments are mostly used as color antioxidants, intensifiers, additives, and pigments are available in different types of colors, they may be most of the time water soluble [3]. Most fungi and bacteria are extensively studied to check potential of the food source colorants. Most of the natural pigments are showing the different activity such as anticancer activity, shows stable towards light, pH, and heat [4]. Therefore the food producers increasingly demanding of bacterial technology to yields colors for used in different foods. Additionally, natural colorants will not a single helpful to the healthiness of humans, but also it will be an advantage for the protection of the environment as destructive chemicals are produced during the developments of different synthetic colorants. These naturally originated colors are widely used breakfasts, baby foods, sauces, pastes, cheese, vitamin containing various milk products, fruits jams and drinks, and some energy beverages [5].

Consequently, natural origin colors, additionally to becoming eco-friendly, can also help for the dual purpose for probiotic health aids and enhancing the external appearance of the product [6]. Few of the significant natural origin pigments are flavonoids, xanthophylls, carotenoids, and tetrapyrroles. The natural pigment most generally used in food manufacturing industries is β-carotene, which is got from cyanobacteria and microalgae. Microbes which having the capacity to give pigments with high yield including the species, like *Serratia*, paecilomyces, Monascus, *Streptomyces, cordyceps,* and yellow-red, blue pigment producing *Penicillium atrovenetum, sarcina, Penicillium herquei, rhodotorula, Monascus purpureus, cryptococcus,* etc., and many more [7]. Among this, *Monascus* species and fungi are widely used in the manufacturing of traditional food like bean curd and different wines. Presently it is required that, the searching for the pigment produced microbes and application of them in commercial point of view. Recently, discovering for the microorganisms producing toxicity free metabolites has been studied by numerous researchers. Furthermore, the waste produced from the food industries can also one of the component for the growth of pigment generating microorganisms. Generally, food industry produces the large amount of waste products such as seeds, peels, rags kernels and pomace, etc. [8, 9]. These are decomposable in nature and widely used as component for the growing of these microbes. These waste product in very rich containing with different minerals, proteins, dietary fibers and carbohydrates, it's a good source as nutrients for this type of microorganisms [8, 10]. Therefore, it is important to explore the different natural bases of a food grade color pigments and there potential benefits. Application of microbial colorants in different foods are shown huge economic potential to explore. The microbial pigments production is presently one of the developing areas of research to establish its potential for different industrial uses [11]. Lots of microbial pigments manufacture is quiet at the elopement stage. Consequently, a study on the microbial pigments should be deepened specifically in discovering inexpensive and appropriate growing medium, which can be decreased the cost and enhance its applicability for industrial manufacturing [12]. The employment of natural pigments in dyes, foods, pharmaceuticals, and cosmetics production methods has enhanced in the recently because of their toxicity free nature. Taken into consideration of this all types of applicability of microbial pigments, it should be necessary to work on the microbial pigments [11].

7.2   FOOD TECHNOLOGY AND COLORS

The present customer's first choice for naturally obtained colorants is accompanying with their motive of being fit. Naturally obtained colorants have becoming extensively famous in customers because of various disadvantages occurs due to synthetic colorants, among some of them showing the intolerance and allergic reaction. Even though, centuries of importance in natural pigments, our awareness regarding its properties, availability, and distribution are so limited. As we know the several drawbacks of marketed natural food colorants such as instable with pH, light, and heat. Persons working with food industries are frequently ask for better substitutes for existing sources of food pigments. It is predicated that the almost 4000 tons of anthocyanins are used each year in the United Kingdom, mainly in foods [13].

The improvement of foods thru a good-looking exterior is a significant objective in the food processing. Progressively, food manufacturers are moving towards natural source food colorants, while certain synthetic color ingredients showing adverse effects in body. Because of the absence of accessibility of natural food pigments, its call is so much specifically in food processing. Existing demand can be accomplished by new research to gives an additional natural well means of pigmented foods and delivers a clean label to it [2]. As a result, it is important to discover different natural sources of food-grade pigments and their capabilities. Presently, Lots of natural pigments are available, among them bacterial pigments play an important role as a food coloring component, since its easy manufacturing and down streaming method. At the industrial level, the manufacturing of natural food pigments via bacterial fermentation has different benefits like easy to extraction, cheaper manufacturing cost, no shortage of raw materials and provides a higher yield in production. Microbes could be genetically altered by introducing desired gene coding for the different pigments, if not produced by microbes. Discussed pigments are found their harmless use as a natural food pigment and it will not only help humans but also it contributes to conservation of biodiversity [14]. Formulation scientists have isolated different food-grade colorants from microbes and blue colorant from the cultured soil microorganisms and could provide natural pigments with an outstanding stability and toxicity profile for food processing. Several scientists are reported that the pigments are trapped into new era of developing food pigments [15]. Most of microbial pigments are previously apply in the fish industries improved the outer appearance of farmed fishes. Additionally, few of natural food pigments have commercially potential for consumption as antioxidants. Therefore, the microbial pigments in additionally to being environment kindly, can also functions the dual purpose for improved outer appearance and probiotic healthiness advantages in foods [16]. The fruitful marketing of colorants obtained from the microbes, in both as nutritional supplements and colorants rest on customer welfare and blooming of the products [11]. Various economic, social, and technical factors have effect on the food processing industries over from many decades and food marked improved vastly with lots of food processing before coming in market for sale as safe and food serving as mother's recipe. Provide the food in the well attractive and fresh is a big challenge in front of many food industrialists and this challenge will be accepted and completed by using different naturally occurring food pigments. The pigment manufacturing industries ambitions to comply with the drink and food production requirement by providing lots of pigments to fulfill all the applications, within standard legislative requirements. Nevertheless, continuous ongoing development in the improvement of handling characteristics and improved stability of the pigments using different formulation technologies, food processing methods, and expansion of completely new pigment [17].

7.3 PIGMENT PRODUCING MICROORGANISMS

A lot of microorganisms such as bacteria, algae, yeast, and molds are producing the pigments. Pigment producing microbes should fulfill some criteria like; should have the ability to survive in changeable temperature, pH, and minerals concentration and should have sensible growth condition. Another one should have the capacity to use vast range of carbon and nitrogen sources and should have sufficient practical pigment yield, as well as should have ability to resist the pathogens and pigments must be easily separated from cell mass. Among them, some of the microorganisms are further discussed in detail [18, 19].

7.3.1 BACTERIA

Bacteria are a significant source of colorants like β-carotene and carotenoids in environment. Lycopene is obtained from the *Streptomyces chrestomyceticus subspecies rubescens* while lutein and zeaxanthin are obtained from *Flavobacterium* is achieving significance. *Canthaxanthin* has been obtained from the *Corynebacterium* species and *Rhodococcus maris* bacterium. The *Canthaxanthin* manufacture has been augmented in a media comprising vitamin B_{12}, minerals, ammonium phosphate, hydrocarbons, and malt extracts, etc. In the culture, broth of *Bacillus substilis* brown pigments are produced and it is by using TLC, it's almost five components are resolved [4, 5, 20].

7.3.2 FUNGI

The different species of fungi, are the conspicuous source of β-carotene, and gather it to infrequent degree. Reproducing of sexual forms caused in a remarkable proliferation in β-carotene. Miscellaneous culture of opposite sex strains typically comprises 5–10 times extra colorants than the single culture. However, β-carotene is the most important colorant manufactured by fungi and lycopene is one more colorant manufactured by *Blakeslea trispora*. The accustomed fermented product *ang-khak,* is obtained by fermenting rice along with *Monascus purpureus*, is applicate as colorant [21]. Colorants, manufactured from immobilized cultures of *Monascus*, are an admixture of different six most important associated colorants and very poor soluble in acid. Although, the yellow pigment isolated from monoscoflavin and ankaflavin, these colorants have been obtained from both liquid and solid cultures of fungi, it gives almost ten-folds more than earlier. For the manufacture of different pigments, the molds are grown-up in the solid culture by using corn, rice, oat, barley, and wheat as substrate [4, 5, 22, 23].

7.3.3   YEAST

Lots of yeasts are rich source of natural bacterial pigments. The cultural compositions play a significant role in manufacturing carotenoid from *P. rhodozyma*. A red pigment, astaxanthin is obtained from animal source, but rarely originates from microbes such as *P. rhodozyma*. The yeasts related to Basidiomycetes have capability to employ the urea, which is very less common especially in Ascomycetous yeasts. Although, developing *P. rhodozyma* as an origin for the pigments of belonging to aquaculture animals, it delivers appropriate color for the aqua culture feedstuffs [4, 5, 24].

7.3.4   ALGAE

Algae also produce the different types of pigments which are widely used in the food industries. The β-carotene manufacturing is preferred in the high salinity and high light intensity. In 1985, the first industry accompanying on track with production and selling of a natural β-carotene. Additional, *Rhodophyta* comprises phycoerythrins and phycocyanins, red microalgae (mutually called phycoerythrins) separately from the leading colorants chlorophyll, it may be blue or red, and they are also found in the *Cryptophyta* and *Cyanophyta* species. Like this, pigments establish a most important fraction of the algae cell protein. Nevertheless, only about 10% of the soluble cell proteins in *Gracelera tikvahiae* are phycobiliproteins, it absorbs some amount of light in visible region [4, 5, 24]. Multiple pigments producing different microbes as shown in Table 7.1.

TABLE 7.1   Multiple Pigments Producing Different Microorganisms

Sr. No	Name of Microorganism	Color Produced by Microorganisms	References
1.	**Bacteria**		[4, 5, 20]
	Bacillus sp.	Brown pigment	
	Pseudomonas sp.	Yellow pigment	
	Achromobacter	Creamy pigment	
	Streptomyces sp.	Blue or red or yellow pigment	
	Corynebacterium michigannise	Creamish or Greyish pigment	
	Brevibacterium sp.	Yellow or orange pigment	
2.	**Algae**		
	Dunaliella salina	Red pigment	[4, 5]
3.	**Yeasts**		
	Rhodotorula sp.	Red pigment	[4, 5, 24]
	Yarrowia lipolytica	Brown pigment	
	Phaffia rhodozyma	Red pigment	
	Cryptococus sp.	Red pigment	
4.	**Molds**		
	Helminthosporium avenae	Bronze pigment	[4, 5, 25]
	H. catenarium	Red pigment	
	H. cynodontis	Bronze pigment	
	H. catenarin	Dark maroon pigment	
	H. gramineum	Red pigment	
	Aspergillus sp.	Red or orange pigment	
	Blakeslea trispora	Creamy pigment	
	Aspergillus glaucus	Dark red pigment	
	P. nalgeovensis	Yellow pigment	
	P. cyclopium	Orange pigment	
	Monascus purpureus	Red or yellow or orange pigment	
5.	**Fungi**		
	Aspergillus galucus	Dark red pigment	[5, 26, 27]
	Fusarium sporotrichioides	Red pigment	
	Monascus roseus	Orange-pink pigment	
	Monascus sp.	Yellow pigment	
	Cordyceps unilateralis	Deep-blood red	

TABLE 7.1 *(Continued)*

Sr. No	Name of Microorganism	Color Produced by Microorganisms	References
	Mucarcircinelloides, Phycomyces blakesleeanus	Yellow-orange pigment	
	Ashbya gossypi	Yellow pigment	
	Paecilomyces sinclairii	Red pigment	
	Pacilomyces farinosus	Red pigment	

7.4 MAJOR MICROBIAL PIGMENTS PRODUCED BY DIFFERENT MICROORGANISMS

7.4.1 RIBOFLAVIN

Riboflavin is a yellow-colored water-soluble pigment manufactured by different microorganisms. It is markedly manufactured by *Ascomycetes Ashbya gossypii*, which is favored due to its elevated yield and better genetic stability [28]. The production of riboflavin was done by three major types, this is strong overproducer, moderate overproducer, and weak overproducer. Among them *Candida flaleri, Ashbya, Saccharomyces cerevisiae, Bacillus subtilis,* and *Eremothicum ashbyii* are the strong producers of it [29]. Riboflavin is mostly used in food industry as chemical addition in pastes, cereals, milk products, and energy drinks and processed cheese, etc. [5].

7.4.2 Β-CAROTENE

β-Carotenoids ranges from the ranges from the orange red to yellow colored pigments. Microbes such as *Micrococcus, Agrobacterium, Serratia, Micrococcus, Streptomyces,* and *Sulfolobus* [30]. It is current trend to use consume the natural products (NPs) like sugarcane molasses, cheese whey, glucose syrup, beet molasses, cellobiose, and peat hydrolysate, etc., for the manufacture of carotenoid [12]. Carotenoid has antioxidant property and has progressive effect in contradiction of some diseases. According to the existing research data, shows mostly using *Mucor circinelloides, Phycomyces blakesleeanus,* and *Blakeslea trispora.* Carotenoid is applied as the additives in orange drinks, vegetable oils, microencapsulated beadlets, various emulsions, and margarine [31].

7.4.3 MONASCUS PIGMENT

These pigments belong to the family of *Ascomycetes* and *Monascaceae* group of microbes. The main four kinds of the stains are isolated for the application in industries are *M. froridanus, M. pilosus, M. rubarand,* and *M. Purpureus.* Monascus provides different colors such as yellow, orange, red to the foodstuffs. Monascus not only provides a good color to the food bust also, but it also gives a pleasant flavor to it. Monascus is commonly used as an additive in different hams, meats, sausages, tofu, and red wines, etc. [32]. Degradation of color is the major problem with this type of the pigment. Recently few studies are reported the stability study of the *Monascus ruber* and it reported as positively [32].

7.4.4 LYCOPENE

Lycopene is a red carotenoid and it is an acyclic isomer of β-carotenoid. It is reported that the cis-isomer of lycopene is highly stable and retains advanced antioxidant property in association to trans-lycopene. Using fungus *Fusarium sporotrichioides* has been tried to yields the antioxidant lycopene and food colorants by means of inexpensive substrate corn filaments [33].

7.4.5 ARPRINK RED

Arprink red is a red colorant manufactured from *Penicillin oxalicum*, which is isolated from soil. The codex Alimentarius Commission has been recommended that the amount permitted to use for the manufacturing of different food products [33, 34] (Table 7.2).

7.5 FACTORS AFFECTING ON PRODUCTION AND STABILITY OF MICROBIAL PIGMENTS

During the manufacturing of the microbial pigments, there are so many other processes carried out, therefore lots of factors are there which directly or indirectly effect on the growth of the microorganisms as well as production of the pigments in a well yield. After production of the natural microbial pigments there are so many physical and chemical factors are there, which are affecting on the stability of the pigments. There are so many actors among them some of discussed in further section.

TABLE 7.2 Microorganisms Producing Pigments and There Categories

Sl. No.	Microorganism	Color	Pigment (Chemical Name)	Category	References
1.	**Bacteria**				
	Xanthomonads oryzae	Yellow	*Xanthomonadin*	Acts against photodamage	[11, 35]
	Staphylococcus areus	Yellow	Zeaxanthin	Antioxidant	
	Flavobacterium sp.				
	Pseudoalteromons sp.	Purple	Violacein	Detoxify ROS	
	Janthinobacterium lividum			Antioxidant	
	Streptomyces sp.	Red	Undecylprodigiosin	Antioxidant	
				antibacterial	
	Corynebacterium michigannise	Blue	Indigoidine	Phaeobacter sp.	
				Antimicrobial	
	Staphylococcus aureus	Golden	Staphyloxanthin	Detoxify ROS	
				Antioxidants	
	Psudomonas sp.	Blue, green	Pyocyanin	Cytotoxicity	
				Pro-inflammatory	
	Alpha Proteobacteria	Red	Heptyprodigiosin	Anti-plasmodial	
	Streptococcus agalactiae	Orange-red	Granadaene	Detoxify ROS	
				Antioxidant	
2.	**Algae**				
	Dunaliella salina	Orange	Beta carotene	Anti-oxidant	[35, 36]
				Anticancer	

TABLE 7.2 *(Continued)*

Sl. No.	Microorganism	Color	Pigment (Chemical Name)	Category	References
	Haematococcus pluvialis	Red	Astaxanthin	Anticancer Antioxidant Photo protectant	
3.	**Yeast**				
	Rhodotorula sp.	Orange, Red	Torularhodin	Antimicrobial Antioxidant	[35, 37]
	Phaffia rhodozyma Dendrorhous Xanthophyllomyces	Red, Pink red	Astaxanthin	Antioxidant Anticancer Photo protectant	
4.	**Archea**				
	Haloferax Alexandrines	Orange	Canthanxanthin	Anticancer Antioxidant	[35]
5.	**Protozoan**				
	Plasmodium sp.	Brown black	Hemozoin	Antioxidant	[34]
6.	**Fungi**				
	Penicillium moxalicum	Red	Anthraquinone	Antifungal Virucidal	[34, 35]
	Fusarium sporotrichioides Blakeslea trispora	Red	Lycopene	Anticancer antioxidant	
	Monascus roseus	Orange-pink	Canthaxanthin	Anticancer Antioxidant	

TABLE 7.2 *(Continued)*

Sl. No.	Microorganism	Color	Pigment (Chemical Name)	Category	References
	Monascus sp.	Yellow	Monascorubramin	Antimicrobial	
				Anticancer	
	Cordyceps unilateralis	Deep-blood red	Nephtoquinone	Anticancer	
				Antibacterial	
	Mucarcircinelloides, Phycomyces blakesleeanus	Yellow-orange	Beta carotene	Anticancer	
				Suppression of cholesterol synthesis	
				Antioxidant	
	Ashbya gossypi	Yellow	Riboflavin	Protection	
				Antioxidant	
				Anticancer	
	Monascus sp.	Yellow	Ankaflavin	Anti-inflammatory	
				Antitumor	
	Monascus sp.	Orange	Rubropunctatin	Anticancer	

7.5.1 TYPE OF FERMENTATION

The solid form of fermentation provides the three times more pigment production than the waterlogged fermentation. For example, in the case of the *M. purpureus*, gives more yield in solid state culture than waterlogged. Other than this, many of factors are affecting such as pH, temperature, agitation, mineral, carbon source, nitrogen source, aeration rate, moisture content, etc., on the production of the different pigments [38].

7.5.2 pH

The pH is also one of most significant factor for the bacteriological pigment manufacture. The pH of the medium is production of the different types of pigments. The pH of the culture media effects on the growing of the microorganisms and type of color manufacture. Optimum pH should be there for the better growth of the microorganisms and the better production yield of the pigments. Slightly alkaline to neutral pH is favorable for the lycopene production although acidic pH is favorable for the β-carotene synthesis [39].

7.5.3 TEMPERATURE

The manufacture of microbial pigments significantly depends on the category of the microbes and temperature are the most important factors for the pigment manufacturing. Temperature for the microorganisms is one of the most significant factors affecting on the growth of microbes and production rate of the pigments, but it depends on the microbe types. The development of *Monascus sp.* requires around 25°C to 28°C for the manufacture of pigments; however, Pseudomonas needs around 35°C to 36°C for its better development and manufacture of enhanced pigment yield. Storage area should be always optimized for a sufficient amount of atmospheric temperature [39].

7.5.4 AERATION RATE

While the Monascus pigments are manufactured from the solid-state fermentation (SSF) process, the bedstead of rice is uninterruptedly aerated by passing gas with required humidity. An involuntary aeration rate of higher than 0.5 L. per min which decreases the manufacture of the biomass and pigments as a significance of water loss from the bedstead. However, highest level of pigments are achieved at forced aeration rates of amongst 0.05 to 0.2 L. per min. during the storage of the different pigments make sure to avoid contaminated air and should provide the fresh air in the storage area.

7.5.5 MINERAL

Different minerals plays a significant role in the production of the various pigments. Zinc immobilizes the development in liquid medium while in solid medium dynamic development and pigmentation was experienced. Optimum level of minerals should be present for the better result in pigments.

7.5.6 PRESENCE OF MOISTURE CONTENT

In solid-state culture medium, the elevated level of pigments are produced by the *Monascus ruber* occurs at around 70% of initial moisture content in substrate which is rice. If moisture content is present as higher amount in pigments then it is not good for the storage of the pigments. It may effect on the stability of the pigments or it may alter the properties of the pigments. Therefore, the moisture content plays an important role in the pigment production and in its storage condition.

7.6 APPLICATIONS

7.6.1 APPLICATION IN FOOD INDUSTRY

Color is dynamic eminence characteristic of foods, and it plays a significant character in sensory and customer acceptance of food product [40]. The application of colors are used from ancient India, but recently with vast globalization in the research, different pigments has been used for enhancing the appearance of products in the food industries as well as pharmaceutical industries [41, 42]. Pigments in foods generate psychological and physiological prospects and approaches that are established by tradition, experience, environment, and education [43].

Natural colorants are widely required in Food industry, because of their lesser good safety and no side effects on body. Some fermentation resultant pigments, like β-carotene from the origin of fungus *Blakeslea trispora* in Europe or pigment from *Monascus* in Asia are nowadays widely applying in food industries [9, 44]. Different pigments deliver a better appearance with nutritive as well as other characteristics, therefore they had great demand in food market. There is an obsession within the consumers in market towards the NPs since of destructive effects of artificial pigments or chemicals. The Monascus red colorant, commonly manufacture by Monascus fermented rice, which increases the organoleptic properties of the food products. There are so many microbial pigments, which are currently undergoing in the research; in future, they will provide a bright scope. Apprehension over the impending toxicity of some artificial pigments have directed to improved importance in colorants obtained from the natural origins [45]. Natural pigments or colorants are obtained from the fauna and flora are supposed to be safe for the reason that of non-cancerous, non-toxic, and its biodegradable property [33, 46].

7.6.2 APPLICATION IN PHARMACEUTICAL INDUSTRY

Furthermost studies considered that the bacteria have exposed the effectiveness and the prospective in clinical claims and their colorants has been applied in treating numerous diseases and they also have assured characteristics. Many pharmaceutical products contains with microbial pigments. Lot of coloring secondary metabolites of the microbes having important potential clinical uses and numerous research works is undergoing regarding to treat the life-threatening diseases and disorders. Pigments may acts as anticancer, antimicrobial, antibiotic, and immunosuppressive as well as antiploriferative and many more. Significant development has been accomplished in the pharmaceutical field, and examinations of bioactive compounds manufactured by these microorganisms are quickly growing. Most of experimentation investigated that the microbes have exposed the effectiveness [47]. Anthocyanin is elaborate in an extensive range of bioactives that distress positively the heath characteristics and reduction in the threat of cancer [48, 49]. The *Serratia* or *Streptomyces* can gives a red matter of pyrrolylpyromethene skeleton, which is one of resulting substances; metacycloprodigiosin, prodigiosin, prodigiosin 25–C, and desmethoxyprodigiosin. These constituents have been recognized for an antimalarial and antibiotic effect, exclusively prodigiosin

25–C gives immunosuppressing action [50]. Immunosuppressing action of prodigiosin was initially designated by some researchers. These researchers disclosed the existence of metacycloprodigiosin and prodigiosinin culture medium of *Serratia* and witnessed discriminating inhibition of polyclonal propagation of T-cells as associated to that of B-cells. Embarrassment of cell propagation as well as introduction of cell death has been witnessed in this cell culture [51].

7.7 FUTURE PROSPECTIVE

Consumers are aware about the application of the natural sources of the different additives in food as well as in pharmaceutical products. Microbial colorant production can be improved in enormous quantity via genetic engineering. Most of the pharmaceutical and food additives are obtained from the microorganisms such as bacteria, yeast, molds, and fungi. Taking advantage of microorganisms for the commercial production is the major significant target of novel development in food additives. There is very crucial requirement for the improvement of fresh strain proficient of elevated yield of pigments exhausting a low-cost substrate with very less downstream procedures. Some of microbial pigments are tremendously sensitive to the temperature, pH, and light deviations, consequently they demonstrate very poor stability and easy to degrade. After fruitful commercialization of microbial pigments as pharmaceutical and food pigments need reserves from private and public stakeholders. This might overwhelmed challenges faced in the industry and may lead to improvement of vigorous technologies for the extraction, manufacture, and purification of new pigments from bacteriological sources. On the supplementary outdated side, the bacterial, microalgae, and fungal multiplicity is nevertheless to be discovered, and the novel tools for the molecular identification should be used.

7.8 CONCLUSION

Modern ever-increasing concern regarding the utilization of natural pigment enormously increasing due to prospective carcinogenicity and teratogenicity of synthetic coloring agents. Nowadays natural pigments widely accepted in every field including food production, textile industries, paper fabrication, agricultural practices, water science, and technology, and in pharmaceutical manufacturing. Various sources are utilized for biocolorant production including fruits, vegetables, roots, and microorganisms. Amongst this *Micrococcus, Bacillus, Rhodotorula, Monascus, Phaffia, Sarcina,* and *Achromobacter microorganisms* potentially produce different pigments. These microbial pigments have number of therapeutic properties including anti-cancerous, immunosuppressive, antibiotic, anti-proliferative, bio-degradability. The use of microbial pigments in processed food is an additional propitious area with huge economic potential. However, microbial pigments offer challenges owing to lower stability, high cost, and discrepancy in shades due to deviations in pH. Enormous production technologies utilized for pigment production including strain development, fermentation, metabolic engineering. Furthermore, toxicology studies of these microbial pigments should be highlighted before their use as natural colorants in food products.

KEYWORDS

- **fermentation**
- **metabolic engineering**
- **microbial pigments**
- **microorganism**
- **pigment**
- **strain development**

REFERENCES

1. De Carvalho, J. C., et al., (2014). *Microbial Pigments, in Biotransformation of Waste Biomass into High Value Biochemicals* (pp. 73–97), Springer.
2. Aberoumand, A., (2011). A review article on edible pigments properties and sources as natural biocolorants in foodstuff and food industry. *World J. Dairy Food Sci., 6*(1), 71–78.
3. Tibor, C., (2007). Liquid chromatography of natural pigments and synthetic dyes. *J. Chromatography Library, 71*, 11–19.
4. Joshi, V., et al., (2003). Microbial Pigments. *Indian Journal of Biotechnology, 2,* p. 362–369.
5. Malik, K., Tokkas, J., & Goyal, S., (2012). Microbial pigments: A review. *Int. J. Microbial. Res. Technol., 1*(4), 361–365.
6. Nagpal, N., Munjal, N., & Chatterjee, S., (2011). Microbial pigments with health benefits-A mini review. *Trends Biosci., 4*(2), 157–160.

7. Kumari, H. M., et al., (2009). Safety evaluation of *Monascus purpureus* red mould rice in albino rats. *Food and Chemical Toxicology, 47*(8), 1739–1746.

8. Heer, K., & Sharma, S., (2017). Microbial pigments as a natural color: A review. *Int. J. Pharm. Sci. Res., 8*(5), 1913–1922.

9. Downham, A., & Collins, P., (2000). Coloring our foods in the last and next millennium. *International Journal of Food Science and Technology, 35*(1), 5–22.

10. Joshi, V., & Attri, D., (2005). *Solid State Fermentation of Apple Pomace for the Production of Value Added Products* (p. 180). Pollution in Urban Industrial Environment.

11. Usman, H., et al., (2017). Bacterial pigments and its significance. *MOJ Bioequiv Availab., 4*(3), 00073.

12. Venil, C. K., Zakaria, Z. A., & Ahmad, W. A., (2013). Bacterial pigments and their applications. *Process Biochemistry, 48*(7), 1065–1079.

13. Wissgott, U., & Bortlik, K., (1996). Prospects for new natural food colorants. *Trends in Food Science and Technology, 7*(9), 298–302.

14. Bener, M., et al., (2010). Polyphenolic contents of natural dyes produced from industrial plants assayed by HPLC and novel spectrophotometric methods. *Industrial Crops and Products, 32*(3), 499–506.

15. Jensen, M. B., et al., (2011). Influence of copigment derived from Tasmania pepper leaf on Davidson's plum anthocyanins. *Journal of Food Science, 76*(3), C447–C453.

16. Sivakumar, V., Vijaeeswarri, J., & Anna, J. L., (2011). Effective natural dye extraction from different plant materials using ultrasound. *Industrial Crops and Products, 33*(1), 116–122.

17. Babitha, S., (2009). *Microbial Pigments, in Biotechnology for Agro-Industrial Residues Utilization* (pp. 147–162) Springer.

18. Dufossé, L., (2006). Microbial production of food grade pigments. *Food Technology and Biotechnology, 44*(3), 313–323.

19. Venil, C. K., & Lakshmanaperumalsamy, P., (2009). An insightful overview on microbial pigment, prodigiosin. *Electron J. Biol., 5*(3), 49–61.

20. Keith, K. E., et al., (2007). *Burkholderia cenocepacia* C5424 produces a pigment with antioxidant properties using a homogentisate intermediate. *Journal of Bacteriology, 189*(24), 9057–9065.

21. Blanc, P., et al., (1994). Pigments of Monascus. *Journal of Food Science, 59*(4), 862–865.

22. Quereshi, S., Pandey, A., & Singh, J., (2010). Optimization of fermentation conditions for red pigment production from *Phomaherbarum* (FGCC# 54) under submerged cultivation. *Journal of Phytology.*

23. Gunasekaran, S., & Poorniammal, R., (2008). Optimization of fermentation conditions for red pigment production from *Penicillium* sp. under submerged cultivation. *African Journal of Biotechnology, 7*(12).

24. Additives, E. P. O. F., & Food, N. S. A. T., (2012). Scientific Opinion on the re-evaluation of mixed carotenes (E 160a (i)) and beta-carotene (E 160a (ii)) as a food additive. *EFSA Journal, 10*(3), 2593.

25. Yoshimura, M., et al., (1975). Production of Monascus-pigment in a submerged culture. *Agricultural and Biological Chemistry, 39*(9), 1789–1795.

26. Sardaryan, E., et al., (2004). Arpink Red-meet a new natural red food colorant of microbial origin. In: Dufossé, L., (ed.), *Pigments in Food, More than Colors* (pp. 207–208). Université de Bretagne Occidentale Publ., Quimper, France.

27. Dufossé, L., (2004). *Pigments in Food, More Than Colors*. Université de Bretagne Occidentale.

28. Stahmann, K. P., Revuelta, J., & Seulberger, H., (2000). *Three* biotechnical processes using *Ashbya gossypii, Candida famata, or Bacillus subtilis* compete with chemical riboflavin production. *Applied Microbiology and Biotechnology, 53*(5), 509–516.

29. Blanc, P. J., (2009). Natural food colorants. *Biotechnology-Volume VII: Fundamentals in Biotechnology, 7*, 137.

30. Mohana, D., Thippeswamy, S., & Abhishe, R., (2013). Antioxidant, antibacterial, andultraviolet-protective properties of carotenoids isolated from *Micrococcus spp. Radiation Protection and Environment, 36*(4), 168–174.

31. Kirti, K., et al., (2014). Colorful world of microbes: Carotenoids and their applications. *Advances in Biology, 2014*.

32. Chung, C. C., Chen, H. H., & Hsieh, P. C., (2008). Optimization of the*Monascus purpureus*fermentation process based on multiple performance characteristics. *Journal of Grey System, 11*(2), 85–96.

33. Nigam, P. S., & Luke, J. S., (2016). Food additives: Production of microbial pigments and their antioxidant properties. *Current Opinion in Food Science, 7*, 93–100.

34. Tuli, H. S., et al., (2015). Microbial pigments as natural color sources: Current trends and future perspectives. *Journal of Food Science and Technology, 52*(8), 4669–4678.

35. Pankaj, V. P., & Kumar, R., (2016). Microbial pigment as a potential natural colorant for contributing to mankind. *Res. Trends Mol. Biol.,* 85–98.

36. Wiltshire, K. H., (2000). Algae and associated pigments of intertidal sediments, new observations and methods. *Limnologica, 30*(2), 205–214.

37. Davoli, P., & Weber, R. W., (2002). Carotenoid pigments from the red mirror yeast, *Sporobolomyces roseus. Mycologist, 16*(3), 102–108.

38. Joshi, V., D. Attri, & N. S. Rana., (2011). Optimization of apple pomace based medium and fermentation conditions for pigment production by Sarcina sp. *Indian Journal of Natural Products and Resources, 2*(4), 421–427.

39. Kumar, A., et al., (2015). Microbial pigments: Production and their applications in various industries. *International Journal of Pharmaceutical, Chemical and Biological Sciences, 5*(1).

40. Fernández-García, E., et al., (2012). Carotenoids bioavailability from foods: From plant pigments to efficient biological activities. *Food Research International, 46*(2), 438–450.

41. Dharmaraj, S., Ashokkumar, B., & Dhevendaran, K., (2009). Food-grade pigments from *Streptomyces sp.* isolated from the marine sponge *Callyspongia diffusa. Food Research International, 42*(4), 487–492.

42. Stich, E., Chaundry, Y., & Schnitter, C., (2002). Color, you eat with your eyes. *Int. Food Ingred., 1*, 6–8.

43. Sanjay, K. R., et al., (2007). Safety evaluation of pigment containing *Aspergillus carbonarius* biomass in albino rats. *Food and Chemical Toxicology, 45*(3), 431–439.

44. Yangilar, F., & Yildiz, P. O., (2016). The Final Development Related Microbial Pigments and the Application in Food Industry. *Journal of Science and Technology, 9*(1), 118–142.

45. Hailei, W., et al., (2011). Improvement of the production of a red pigmentin *Penicillium sp.* HSD07B synthesized during co-culture with *Candida tropicalis. Bioresource Technology, 102*(10), 6082–6087.

46. Silveira, S. T., Daroit, D. J., & Brandelli, A., (2008). Pigment production by *Monascus purpureus* in grape waste using factorial design. *LWT-Food Science and Technology, 41*(1), 170–174.
47. Soliev, A. B., Hosokawa, K., & Enomoto, K., (2011). Bioactive pigments from marine bacteria: Applications and physiological roles. *Evidence-Based Complementary and Alternative Medicine, 2011.*
48. Lazzè, M. C., et al., (2004). Anthocyanins induce cell cycle perturbations and apoptosis in different human cell lines. *Carcinogenesis, 25*(8), 1427–1433.
49. Kong, J. M., et al., (2003). Analysis and biological activities of anthocyanins. *Phytochemistry, 64*(5), p. 923–933.
50. Tirumale, S., & Wani, N. A., (2018). Biopigments: Fungal Pigments. In: *Fungi and their Role in Sustainable Development: Current Perspectives* (pp. 413–426). Springe.
51. Pandey, R., Chanderb, R., & Sainisc, K. B., (2007). Prodigiosins: A novel family of immunosuppressants with anti-cancer activity. *Indian Journal of Biochemistry & Biophysics. 44*, 295–302.

CHAPTER 8

Ethnopharmacological Perspectives of the Traditional Herb *Tabermontana divaricate Linn.*

SEEMA WAKODKAR,[1] DEBARSHI KAR MAHAPATRA,[2] SHILPA BORKAR,[1] and SHAILAJA LATKAR[1]

[1]*Department of Pharmacology, Kamla Nehru College of Pharmacy, Nagpur – 441108, Maharashtra, India,*
E-mail: seemausare@rediffmail.com (S. Wakodkar)

[2]*Department of Pharmaceutical Chemistry, Dadasaheb Balpande College of Pharmacy, Nagpur – 440037, Maharashtra, India*

8.1 INTRODUCTION

Herbals are the natural remedies against human diseases and find applications for centuries because they restrain variable components of therapeutic value; some are also used for a preventive reason. Across the globe, rising attention towards the herbal remedies has been included in the conventional medicinal plant practice. Traditional uses of medicinal plants cope up with diseases such as infantile convulsion, diarrhea, dysentery, malaria, epilepsy, fungal, and bacterial infections [1]. Herbal remedies are believed to be a "chemical factory" due to an abundance of chemical compounds like oleoresins, sesquiterpene, lactones alkaloids, glycosides, saponins, resins, and oils (essential and fixed) [2]. India acknowledged as one of the oldest, richest, and most diverse custom attitudes associated with the use of medicinal plants. The World Health Organization (WHO) encourages the efficacious, safe, less toxic, accessible, and reliable use of herbals [3]. Nowadays, there is mounting awareness in the chemical opus of plant-based medicines. Bioactive elements have been isolated and studied for pharmacological screening. During the last two decades, the pharmaceutical industry has made enormous speculation in chemical and pharmacological explorations where the entire globe puts effort to find out more potent drugs and a few new ones. The commercial screenings test plants have successfully passed [4].

The genus *Tabemaemontana* was named after J. Th. Miller, a German physician, and botanist who was born in Bergzabem and died in Heidelbergin the Pfalz in 1690; he Latinized his birthplace as *Tabemaemontanus*. Genus, belonging to the Apocynaceae and comprising about 100 species are distributed throughout the world. *T. divaricata* (L.) is widely located throughout India as an ornamental plant. It possesses comprehensive activities like anti-cancer, anti-infection, anti-convulsant, anti-oxidant, anti-diabetic, anti-inflammation, and anthelmintic activity properties [5–10].

- **Botanical name**: *Tabernaemontana divaricata*
- **Family**: Apocynaceae
- **Common name**: Wax flower, Kathana, Crepe jasmine, Crepe flower
- **Part used**: Root, stem, leaves, and flowers
- **Habitant**: Grows as a common weed.

In India, it occurs in Khasia Hills, upper Gangetic plain, East Bengal, Assam, Garhwal, hills of Vishakapatnam, and Burma. It is cultivated as an ornamental plant, grows wild in hedges and shady forests [11].

8.2   MORPHOLOGY

8.2.1   PHYTOCHEMICAL CONSTITUENTS ISOLATED FROM TABERNAEMONTANA DIVARICATA

Phytochemical constituents isolated from *T. divaricata*: The phytochemical analysis showed the presence of flavonoids, alkaloids, lignins, glycosides, tannins, saponins, amino acids, quinines, and carbohydrates. The phyto-chemical analysis of stem bark extracts in petroleum ether, chloroform, methanol, and aqueous solvents of indigenous medicinally important plants of *T. divaricata* were investigated. Everywhere, its species are used in traditional medicine (TM) and for other purposes. In the chemical screening of *Tabernaemontana* species, usually alkaloids are found and only and occasionally other important secondary plant metabolites. Variety of alkaloids has been isolated from the plant bark of the stem and root including coronaridine, tabernaemontanine, coronarine, and dregamine. So the alkaloids are referred to as a key factor for most of the phytochemical work. Alkaloids are situated in all the vegetative parts of the shrub. The applications are anxiolytic, antifertility, antibacterial, antiobesity, hypolipidemic, anticonvulsant, anticancer, antinociceptive, antidiarrheal, antidiabetic, cytotoxic, gastroprotective, antioxidant, analgesic, and anti-inflammatory. *T. divaricata* flower juice mixture with oil attenuates burning sensation, cures sores of eye and skin diseases, inflammation can be prevented with the application of leaves juice to wounds and used in opthalmia [12]. Antimicrobial action against infectious diseases such as syphilis, gonorrhea, and leprosy, as well as parasitic action against worms, dysentery, diarrhea, and malaria, is the most common use of *T. divaricata* extract [13] (Tables 8.1 and 8.2).

TABLE 8.1   Molecular Weight and Various Components Identified in Leaves

Sl. No.	Name of the Compound	Molecular Formula	Molecular Weight
1.	3,7,11,15-Tetramethyl-2-hexadecen-1-ol	$C_{20}H_{40}O$	296
2.	Lactose	$C_{12}H_{22}O$	342
3.	n-Hexadecanoic acid	$C_{16}H_{32}O_2$	256
4.	Phytol	$C_{20}H_{40}O$	296
5.	9,12-Octadecadienoic acid (Z,Z)-	$C_{18}H_{32}O_2$	280
6.	Squalene	$C_{30}H_{50}$	410
7.	Cedrol	$C_{15}H_{26}O$	222
8.	Vitamin E	$C_{29}H_{50}O_2$	430
9.	Urs-12-en-24-oic acid, 3-oxo-, methyl ester, (+)-	$C_{31}H_{48}O_3$	468

TABLE 8.2 Phytoconstituents Present in the Ethanolic Leaf Extract of *T. divaricata* L.

Sl. No.	Name of the Compound	Activity
1.	3,7,11,15-Tetramethyl-2-hexadecen-1-ol	Antimicrobial, anti-inflammatory
2.	Lactose	Preservative, nutritive
3.	n-Hexadecanoic acid	Antioxidant, pesticide, anti-androgenic
4.	Phytol	Antimicrobial, Anticancer, Anti-inflammatory
5.	9,12-Octadecadienoic acid (Z,Z)	Hypocholesterolemic, cancer preventive, hepatoprotective nematicide, anti-acne, anti-arthritic
6.	Squalene	Anticancer, antimicrobial, antioxidant, pesticide
7.	Cedrol	Anti-tumor, antibacterial, anti-inflammatory, fungicide, nematicide
8.	Vitamin E	Anti-aging, anti-diabetic, antioxidant, anti-leukemic, hepatoprotective,anti-ulcerogenic, anti-bronchitic
9.	Urs-12-en-24-oic acid, 3-oxo methyl ester, (+)	No activity reported

8.2.2   PHARMACOLOGICAL AND BIOLOGICAL ACTIVITIES OF T. DIVARICATALINN

8.2.2.1   ANALGESIC AND ANTI-INFLAMMATORY ACTIVITY

The ethanolic extract of *T. divaricata* L. flowers was evaluated for analgesic and anti-inflammatory effects on the Wistar rat model. The anti-inflammatory and analgesic potential of the flower extract at doses of 125 mg/kg, 250 mg/kg, and 500 mg/kg was screened against indomethacin (20 mg/kg p.o.), the standard drug. Wistar rats of either sex were employed for the study and evaluated by using hot plate reaction time, acetic acid-induced writhing, safety test on gastric mucosa method, and carrageenan-induced hind paw edema. The ethanolic extract presented an anti-nociceptive activity in acetic acid-induced writhing as distinguished by a momentous lessening in the number of writhing in rats ($p < 0.01$). In the hot plate test, the extract demonstrated potential nociceptive effect towards thermal stimuli in the animal subjects, and a considerable enhancement in the reaction time was viewed ($p < 0.01$). The test drug appreciably restrained the carrageenan-induced hind paw edema in rats which represented the anti-inflammatory effect (p < 0.01). Though, no gastric lesions were seen in extracted-treated rats which pointed out towards the safety limit. The ethanolic extract presented the potential as anti-inflammatory and analgesic agents in different animal models [12].

8.2.2.2 ANXIOLYTIC ACTIVITY

The present study was conducted to access the anxiolytic activity of alcoholic flower extract of *T. divaricata* (Linn) R.Br. (ALETD) by using mice. This plant contains amino acids, carbohydrates, proteins, tannins, flavonoids, triterpenes, glycosides, and steroids. The anxiolytic effect of ALETD (100, 200, and 400 mg/kg) was studied and diazepam used as a standard drug by using following animal models; light-dark transition test (LDT), open-field test (OFT), and elevated plus maze (EPM). On EPM, all the doses of extract and the diazepam had exhibited important anxiolytic activity by growing open arm entries and time spent in open arm. In OFT, enhanced central locomotion, total locomotion, and reduced frequency of rearings, immobility time was perceived with all the doses of the extract and diazepam. In LDT model, the anxiolytic activity was predominantly seen with diazepam as well as all the doses of the extract by frequency of crossings, escalating latency, decreasing rearings in light box, and time spent in light box were scrutinized. The flavonoids, alkaloids, and other chemical constituents present in ALETD are contemplated to relate with the monitored pharmacological activities of the plant's extract in the experimental animal model employed. The discovery from this animal study delegated that ALETD possesses the anxiolytic property and thus provide pharmacological credibility to the ethnomedical and folkloric applications of the plant in the pharmacotherapeutics of anxiety condition [13].

8.2.2.3 ANTI-FERTILITY ACTIVITY

The study aimed at the anti-fertility screening of *T. divaricata* leaf extracts continually administered to the male rats. The male rats focused to the effect of 50% ethanol extract of *T. divaricata* leaves on reproduction. Twenty animals divided into four groups for study. The first groups (I) serve as control received vehicle alone. For the critical period of 60 days, the II, III, and IV groups were administered with the leaf extract every day at 50 mg/kg, 100 mg/kg, and 200 mg/kg body weight p.o., respectively. A drastic decrease in the mass of epididymis, testes, ventral prostate, and seminal vesicle was detected. A dose-related diminution in the epididymal sperm count, sperm motility, testicular sperm count, pups population, the ratios between the delivered females and inseminated females, and the number of the fertile males were inspected. A clear interdependence observed between the dose and severity of lesions of seminiferous epithelium in testis. In general, the seminiferous tubules come out minimum in size with a recurrently-filled eosinophilic material. Spermatogenesis gets arrested at the secondary spermatocyte stage. Leydig cells were seen atrophied. Morphological changes were not observed in Sertoli cells. The serum concentration of luteinizing hormone and testosterone were significantly reduced. In serum FSH concentration, no marked change was recorded. Finally marked elevation in body weights of all groups were. The hematological parameters remain unaltered. The conclusion for the study that the 50% ethanol extract of *T. divaricate* leaf produced dose-dependent effect on male reproduction without altering general body metabolism [14].

8.2.2.4 ANTI-BACTERIAL ACTIVITY

T. divaricate L. on physiochemical analysis of the dried leaves indicated the presence of a steroid, reducing sugar, saponins, tannins, and gums. The current investigation was focused on the evaluation of anti-bacterial activity of *T. divaricate* (L). The maximum potency against infectious pathogens; *Staphylococcus pyogenes*, *S. saprophyticus*, *S. aureus*, *Escherichia coli*, *Enterococcus facealis*, *Streptococcus agalactae*, *Shigella boydii*, *S. dysenteriae*, *Salmonella typhi*, and *Pseudomonas aeruginosa* were perceived. The zone of inhibition (ZOI) was observed with roughly all bacteria with some omission, the minimum inhibitory concentrations (MICs) of the screened extracts were found to be noteworthy. The obtained results provided underpin for the use of this plant in TM and its additional exploration [15].

8.2.2.5 GASTROPROTECTIVE EFFECT

The present study evaluates the anti-ulcer attributes of methanolic extract of *T. divaricata* flower (TDFME 500 mg/kg, p.o.) by gastric ulceration-induced pyloric ligation model utilizing omeprazole (8 mg/kg, p.o.) as a standard drug in Wistar rats. The five parameters like pH, volume of gastric juice, ulcer index, and free and total acidities were assessed. The test drug extracts significantly (p< 0.01) lessen the above parameters drastically. As standard, it also elevated pH of gastric acid. The % protection for the test and the standard were originated to be 79.53% and 89.84%, respectively. TDFME in 500 mg/kg dose had a positive effect overall the strictures under the study and the results were analogous with the standard. The above-mentioned results concluded that TDFME expressed a notable gastroprotective effect [16].

8.2.2.6 ANTI-DIABETIC AND CYTOTOXIC ACTIVITIES

The examinations reported anti-diabetic perspectives of *T. divaricata* methanolic floral extract in alloxan-induced diabetic mice. The extract was administered i.p. at a single dose of 200 mg/kg and 300 mg/kg b.w. and the blood glucose levels were measured at an interval of 0 hr, 1 hr, 3 hrs, 5 hrs, 10 hrs, and 24 hrs of the study period. The anti-hyperglycemic effect of the extract was correlated with the standard drug metformin. The dose found to be more effective dose was 300 mg/kg. This dose reduces the elevated blood glucose level from 14.15 ± 0.42 mg/dL to 8.81 ± 0.27 mg/dL observed at 10th hr of the treatment period, whereas extreme result was obtained for metformin at the same time from 14.04 ± 0.36 mg/dL to 6.13 ± 0.19 mg/dL. The extract was also focused on Brine shrimp lethality bioassay. The LC_{50} value of the screened extract was expressed to be 84.03 µg/mL and for vincristine sulfate, it was observed to be 10.58 µg/mL. So, the above-mentioned results suggest that *T. divaricata* possess anti-diabetic activity in alloxan-induced diabetic mice and low cytotoxicity that opened emerging pharmacotherapeutic approaches towards diabetes [17].

8.2.2.7 ANTICONVULSANT ACTIVITY

The investigation was planned to determine the spontaneous motor activity (SMA) and anticonvulsant potential of alcoholic extract of *T. divericata* flowers (ALETD) by employing diverse exemplars. The anticonvulsant properties were tested on animal models such as isoniazid (INH)-induced convulsion (IIC), maximal electroshock-induced convulsion (MESIC), pentylenetetrazole (PTZ) induced convulsions (PTZIC), 4-amino pyridine (4-AP)-induced convulsion (4-APIC), picrotoxin-induced convulsion (PIC), and strychnine-induced convulsions (SIC) in mice. The level of Gama amino butyric acid was also approximated. A significant decline in the locomotor score was evidenced with diazepam (5 mg/kg) and ALETD (100, 200, and

400 mg/kg) which were observed with anticonvulsant activity by diminishing the period of tonic extensor phase in the experimental animals. In PIC, SIC, IIC, PTZIC, and 4-AMIC models, a diazepam and in all dose level of ALETD offered an amplified onset of tonic convulsion. In the mice serum amplified GABA level was observed with gabapentin (20 mg/kg) and extract. The flavonoids, alkaloids, and miscellaneous chemical constituents were detected in ALETD are contemplated to relate for the examining pharmacological consequences of the plant's extract in the animal models archetypes employed [18].

8.2.2.8 ANTI-OBESITY AND HYPOLIPIDEMIC ACTIVITY

T. divaricata was traditionally used to treat the cases of circulatory irregularities and arteriosclerosis. The research was frame worked to assess the probable consequence of methanolic extract of T. divericata aerial parts (METD) on atherogenic diet-induced obese rats on obesity and hyperlipidemia. METD for a period of 42 days was administered at the doses of 100 mg/kg and 200 mg/kg along with atherogenic diet in rats. The parameters evaluated were serum lipid profiles, body temperature, body weight, SGOT, and SGPT. As that of standard drug sibutramine (2 mg/kg) the altered parameters are corrected by METD at the dose of 200 mg/kg significantly ($p<0.001$). T. divaricata exerted noteworthy anti-obese and anti-hyperlipidemic effects in rats fed on atherogenic diet. It could be envisaged that from the remarks of the executed study. Further examinations are needed to understand the pharmacotherapeutic perspectives and precise mechanism of T. divaricate as an anti-obesity drug [19].

8.2.2.9 ANTI-DIARRHEAL ACTIVITY

Studies were subjected to investigate anti-diarrheal activity of T. divaricata leaves aqueous and hydroalcoholic extracts on castor oil-induced diarrhea and gastrointestinal (GI) motility in rats. The dose-dependent protection against castor oil-induced diarrhea and a marked decrease in GI motility were observed with the hydroalcoholic and aqueous extracts of T. divaricata leaves (100, 200, and 300 mg/kg, p.o.). A preliminary phytochemical screening reveals the presence of tannins, alkaloids, proteins, resins, flavonoids, amino acids, phenols, saponins, steroids, glycosides, triterpenoids, fixed oils, and fats in extracts of T. divaricata leaves. The results obtained showed significant activity against diarrhea of hydroalcoholic and aqueous extracts of *T. divaricata* leaves, so it can be used traditionally for GI disorders [20].

8.2.2.10 ANTI-NOCICEPTIVE ACTIVITY

T. divaricata L. commonly known as Togor in Bengali, Dudhphul in Bangladesh, Waxflower, or Crepe jasmine worldwide, is an evergreen shrub to 6 feet (1.8 m) distributed in the coastal forests of Myanmar, Bengal, mangrove forests of China and Japan. *T. divaricata* extract inhibited the neuronal acetylcholinesterase activity in the animal model [22]. Although, a number of chemical investigations have been done and several constituents have been found such as glycosides, carotenoids, flavonoids, triterpenes, polyphenols, alkaloids, lipids, saponins, etc. [23–25].

8.2.2.11 ANTI-CANCER ACTIVITY

The study evaluates anti-cancer activity of the aqua-alcoholic extract of the flowers of *T. divaricata*, traditionally used for the treatment of a number of diseases. The soxhlet extraction performed with petroleum ether and hydroalcohol. The anticancer study was accomplished *in-vitro* against the human cervical cancer cell line (HeLa) and MTT assay was used to analyze the cell growth inhibition. It was confirmed from the results that the hydroalcoholic extract of flowers possessed a moderate anticancer activity and the IC_{50} value was greater than 100 µg/mL [26].

8.2.2.12 ANTI-CATALEPSY ACTIVITY

The study evaluates, haloperidol (1.0 mg/kg, i.p.) induced significant catalepsy in rats. The study revealed that both the extracts viz.; ethanolic and aqueous extract of *T. divaricata* protect rats from catalepsy as compared to the standard drug, levodopa. In the above study, it was observed that the various extracts reduce the duration of catalepsy in rats. The phytochemical screening of *T. divaricata* evaluated the presence of steroids, proteins, saponins, tannins, amino acids, alkaloids, phenols, glycosides, resins, triterpenoids, fixed oils, flavonoids, and fats. The cataleptic score significantly reduced at 90 min (p<0.001) with ethanolic and aqueous extract (50–150 mg/kg, i.p.). The activities at 50–150 mg/kg dose i.p. are comparable with that of standard levodopa [27].

8.2.2.13   TOXICITY

Heteropneustes fossilis (Bloch) were used to study toxicological effects of plant parts extracts (dry leaf, bark, and fruit) of *T. divaricata* L. The extracts of plant parts of *T. divaricata* were prepared by using distilled water, hydroalcoholic, and 100% ethyl alcohol and under laboratory conditions tested on the experimental fishes. The percentage mortality depends upon the type of plant extracts and concentrations, which varied between 13.33–93.33%. The extracts were presented for the period of 24 hrs exposure period to evaluate mortality (%), LC_{50}, LC_{90}, and F-values. The mortality was concentration-dependent. The toxicity was observed as 100% ethyl alcohol > hydroalcoholic >distilled water for plant extracts. LC_{50} values gave relative toxicity of the pesticides on *H. fossilis* was found to be *T. divaricata* fruit extract > bark extract > leaf extract [28].

8.2.2.14   ROLE OF OTHER SPECIES OF TABERNAEMONTANA AS ANTI-INFLAMMATORY

The progression of ear edema was induced by phenol, arachidonic acid, cinnamaldehyde, croton oil, and capsaicin with a maximum inhibition of 100% to phenol, croton oil, arachidonic acid, and cinnamaldehyde and $75 \pm 6\%$ to capsaicin was prevented by *T. catharinensis* TcE as an anti-inflammatory agent (10 μg/ear). Besides, the TcE (10 μg/ear) also prevented the augmentation of MPO activity to the same irritant agents. The positive controls also averted both ear edema and the enhancement in MPO activity, i.e., indomethacin or dexamethasone in skin inflammation model (induced by phenol) and dexamethasone in inflammation model (induced by croton oil). TcE also leads to an increase of MIP-2, IL-1β, and TNF-α level and prevented the inflammatory infiltration irritant agents-induced. Terpenes, phenolic compounds, and indole alkaloids may be a characteristic component of TcE topical anti-inflammatory effect to the combined effect of extract and recognized by the de-replication method. In preliminary toxicological tests, the TcE's therapeutic dose proved to be a safe and result suggests attractive approaches for the management of certain inflammatory diseases [29].

8.2.2.15 TOXICOLOGICAL EVALUATIONS OF THE HYDROETHANOL EXTRACT OF T. CRASSASTEM BARK

A variety of affections are treated by *T. crassa* Benth is a medicinally privileged plant that is widely used in folk medicine. The evaluation of its toxicological profile via histopathology results showed a dose-related effect in the lungs, liver, and kidneys, no difference occurred in the tissue profile of those receiving 6 weeks daily treatment of 0.5 g/kg b.w. and the control group. The acute toxicity results indicated the LD_{50} value of 6.75 g/kg body weight (b.w.) after 48 hr of treatment and the considerable variation (p<0.05) of the relative body weight, total bilirubin (TBil), serum alkaline phosphatase (ALP), direct bilirubin (DBil), creatinine (SCr), and alanine aminotransferase (ALT), at a dose of 6 g/kg b.w. The significant variation also occurred in the total proteins (TP), aspartate aminotransferase (AST), liver ALP, glutathione (GSH), ALT, malondialdehyde (MDA), urea (RU), and renal creatinine (RCr), at a dose of 6 g/kg b.w. The important changes in the body weight were perceived in a sub-acute toxicity study but no variation in (P>0.05) of liver indices and blood for the animal taking 6-weeks daily doses of the HE as compared to the control group. The consumption of up to 6 g/kg b.w. higher doses could cause injuries to the organ. A moderated consumption of 0.5 g/kg b.w. daily for 6-weeks small doses appeared to be safe [30].

8.3 CONCLUSION

Thousands of constituents from plants possessing bioactivities are the precious foundation of novel and biologically-active compounds. The herbal extracts present in many products are as supplements or substitutes of present-day medicine sold in the Asian market. In recent years, traditional, and ethnobotanical uses of natural compounds, of plant origin obtain a lot of consideration as they are well tested for their effectiveness and usually supposed to be safe for human use. The book chapter has been aimed to illustrate the latest advancements in the examination of plant *T. divaricata* as phytotherapy and to illustrate its potential as a therapeutic agent. The identification, cataloging, and certification of plants necessitate comprehensive and methodical study that may provide a consequential means for the encouragement of the traditional knowledge of the herbal medicinal plants. The conclusion of the study is that plant *T. divaricata* having an extensive range of medicinal value due to an ample variety of phytoconstituents can be additional examined based on other parameters to obtain a valuable market product.

KEYWORDS

- **bioactivities**
- **ethnomedicine**
- **folk medicine**
- **pharmacology**
- **phytochemical**
- ***Tabermontana divaricate***

REFERENCES

1. Sofowora, A., (1996). *J. Altern. Complement. Med., 2*(3), 365–372.
2. Singh, A. P., (2005). Promising phytochemicals from Indian medicinal plants. *Ethnobotanicals Leaflets, 2005*(1), Article 18.
3. Kalaimagal, C., & Umamaheswari, G., (*2015).* Bioactive compounds from the leaves of *Tabernaemontana divaricata* (L).*International Journal of Recent Scientific Research, 6*(4), 3520–3522.
4. Rahman, M., Ashikur, M., Hasanuzzaman, R. M., Mofizur, Shahid, I. Z., & Roy, S. M., (2011). Evaluation of antibacterial activity of study of leaves of *Tabernaemontana divaricata* (L). *International Research Journal of Pharmacy, 2*(6), 123–127.
5. Nowshin, N. R., Mostafizur, R. M., & Khalequzzaman, K. M., (2012). Antioxidant and cytotoxic potential of methanol extract of *Tabernaemontana divaricata* leaves. *International Current Pharmaceutical Journal, 1*(2), 27–31.
6. Qamruzzama, Javed, A. A., & Mateen, S., (2012). Analgesic and anti-inflammatory effect of ethanolic extract of *Tabernaemontana divaricata* L. flowers in rats. *Der. Pharmacia. Lettre., 4*(5), 1518–1522.
7. Hari, B. V., Dantu, A. S., Shankarguru, P., & Devi, D. R., (2012). Qualitative Analysis and Anthelmintic Activity of Hydro-Alcoholic Extract of *Tabernaemontana divaricata*. *Asian Journal of Pharmaceutical Research and Health Care, 4*(2), 63–69.
8. Basavaraj, P., Shivakumar, B., Shivakumar, H., & Manjunath, V. J., (2011). Evaluation of Anticonvulsant activity of *Tabernaemontana divaricata* (Linn) R. Br. Flower Extract. *International Journal of Pharmacy andPharmaceutical Sciences, 3*(3).
9. Kirthikar, K. R., & Basu, B. D., (1998). *Indian Medicinal Plants. Vol.III.* Dehradun: Bishen Singh Mahendra Pal Singh, 577–78.

10. Shazid, M. S., Samabesh, C., & Ahmed, A. R., (2011). Phytochemical and antinociceptive study of leaves of *Tabernaemontana divaricata* (L). *Journal of Medicinal Plants Research, 5*(2), 245–247, ISSN 1996-0875.

11. Akhila, S. D., Shankarguru, P., Ramya, D. D., & Vedha, H. B., (2012). Evaluation of anticancer activity of hydroalcoholic extract of *Tabernaemontana divaricata*. *Asian Journal of Pharmaceutical and Clinical Research, 5*(4), ISSN-0974-2441.

12. Qamruzzamaa, Javed, A. A., & Mateen, S., (2012). Analgesic and anti-inflammatory effect of ethanolic extract of *Tabernaemontana divaricata* L. flowers in rats. *Der Pharmacia Letter, 4*(5), 1518–1522, ISSN: 0975-5071, USA CODEN: DPLEB4.

13. Basavaraj, B. S., & Shivakumar, H., (2011). Anxiolytic activity of *Tabernaemontana divaricata* (Linn) R. Br. flowers extract in mice. *International Journal of Pharma. and BioSciences, 2*(3). ISSN: 0975-6299.

14. Jain, S., Jain, A., Paliwal, P., & Solanki, S. S., (2012). Antifertility effect of chronically administered Tabernaemontana divaricata leaf extract on male rats. *Asian Pacific Journal of Tropical Medicine, 5*(7), 547–551.

15. Rahman, M., Ashikur, M., Hasanuzzaman, Rahman, M. M., Shahid, I. Z., & Roy, S. M., (2011). Evaluation of antibacterial activity of study of leaves of *Tabermontanadivaricata*(L). In: Rahman, M., Ashikur, et al., (eds.), *IRJP* (Vol. 2, No. 6, pp. 123–127). ISSN: 2230-8407.

16. Mohammed, S. A. K., (2011). Gastro protective effect of *Tabernaemontana divaricata* (Linn.) R.Br. flower methanolic extract in wistar rats. *British Journal of Pharmaceutical Research,* (3), 88–98.

17. Masudur, R. M., Saiful, I. M., Sekendar, A. M., Rafikul, I. M., & Zakir, H. M., (2011). Antidiabetic and cytotoxic activities of methanolic extract of *Tabernaemontana divaricata* (L.) flowers. *International Journal of Drug Development and Research, 3*(3), ISSN: 0975-9344.

18. Basavaraj, P., Shivakumar, B., Shivakumar, H., & Manjunath, V. J., (2011). Evaluation of anticonvulsant activity of *Tabernaemontana divaricata* (Linn) R. Br. flower extract. *International Journal of Pharmacy and Pharmaceutical Sciences, 3*(3), ISSN: 0975-1491.

19. Kanthlal, S. K., Suresh, V., Arunachalam, G., Royal, F. P., & Kameshwaran, S., (2012). Anti-obesity and hypolipidemic activity of methanolic extract of *Tabernaemontana divaricata* on atherogenic diet induced obesity in rats. In: Kanthlal, S. K., et al., (eds.), *IRJP* (Vol. 3, No. 3). ISSN: 2230-8407.

20. Chanchal, N. R., Balasubramaniam, A., & Sayyed, N., (2013). Antidiarrheal potential of *Tabernaemontana divaricate* (Vol. 4, No. 1, pp. 61–68). Inforesights Publishing.

21. Wasana, P., Anchalee, P., Nipon, C., & Siriporn, C., (2008). Ethnobotany and ethnopharmacology of *Tabernaemontana divaricata*. *Indian J. Med. Res., 127*, 317–335.

22. Basak, U. C., Das, A. B., & Das, P., (1996). Chlorophylls, carotenoids, proteins, and secondary metabolites in leaves of 14 species of mangrove. *Bulletin of Marine Sciences, 58*(3), 654–659.

23. Ghosh, et al., (1985). Extraction and quantification of pigments from Indian traditional medicinal plants: A comparative study between tree. *Shrub and Herb International Journal of Pharmaceutical Sciences and Research, 9*(7), 3052–3059.

24. Sharma, P., Mehta, P. M., & Varansi, (1969). The Chowkhamba Vidyabhawan Dravyaguna Vignyan. Parts II and III; p. 586.

25. Shazid, M. S., Samabesh, C., & Ahmed, A. R., (2011). Phytochemical and antinociceptive study of leaves of *Tabernaemontana divaricata* (L). *Journal of Medicinal Plants Research, 5*(2), 245–247.

26. Akhila, S. D., Shankarguru, P., Ramya, D. D., & Vedha, H. B., (2012). Evaluation of *in-vitro* anticancer activity of hydroalcoholic extract of *Tabernaemontana divaricata*. *Asian Journal of Pharmaceutical and Clinical Research, 5*(4), ISSN: 0974-2441.

27. Raj, et al., (2014). Effect of various extracts of *Tabernaemontana divaricata*on haloperidol induced catalepsy in rats. *International Current Pharmaceutical Journal, 3*(3), 240–242.

28. Ara, S. I., Nasiruddin, M., & Hossain, A., (2012). Toxicity of leaf, bark, and fruit extracts of *Tabernaemontana divaricata* (L.) against *Heteropneustes fossilis* (Bloch). *Bangladesh J. Environ. Sci., 23*, 39–44, @ BAED ISSN 1561-9206.

29. Camponogara, C., Casoti, R., Brusco, I., Piana, M., Boligon, A. A., Cabrini, D. A., Trevisan, G., et al., (2019). *Tabernaemontana catharinensis* leaves exhibit topical anti-inflammatory activity without causing toxicity. *J. Ethnopharmacol., 231*, 205–216.

30. Kuete, V., Manfouo, R. N., & Beng, V. P., (2010). Toxicological evaluation of the hydroethanol extract of *Tabernaemontana crassa* (Apocynaceae) stem bark. *J. Ethnopharmacol., 130*(3), 470–476.

CHAPTER 9

A Closer View on Various Reported Therapeutically Active Formulations Containing Aloe vera *(Aloe barbadensis)*

VAIBHAV SHENDE and DEBARSHI KAR MAHAPATRA

Department of Pharmaceutics, Gurunanak College of Pharmacy and Technical Institute, Nagpur – 440026, Maharashtra India, E-mail: mahapatradebarshi@gmail.com (D. K. Mahapatra)

9.1 INTRODUCTION

Aloe vera is understood as a miracle plant. The foremost best-known species of Aloe plant (i.e., mature worldwide) is succulent *Aloe barbadensis* Miller which belongs to the lily family (Family: Liliaceae). The produced glutinous gel created from the center (parenchyma) of the plant leaf is the most popular product in use and it has immense applications like burn gel commercially [1]. The gel stimulates cell growth and enhances the restoration of broken skin. It moisturizes the skin as a result of its water-holding capability. On drinking, it protects the secretory membrane of the abdomen, particularly once irritated or damage [2]. Dental sickness is a significant illness throughout the planet and therefore the succulent herbs are used. The exceedingly varied pharmaceutical preparations are one amongst the decay issues where it is going to be acute or chronic and treatment needs a spare drug concentration at the specific position with no unwanted effects [3]. Aloe plant on oral administration shows wound healing improvement within the early part along with a single dose acute radiation exposure which up wounds the activity by having a stimulating impact on increase inflammatory cell infiltration, fibroblast proliferation, growth, and protein production [4]. The nanoparticles of Aloe vera showed targeted delivery. The engineering platforms may function customizable where the targeted drug delivery vehicles that are competent of hauling giant quantity of therapeutic agent into the malignant cells whereas selectively avoiding the healthy cells [5]. The artificial anti-microbial agent showed drawbacks of drug resistance and alternative aspect results. In the pharmaceutical world, the gel is the most convenient and patient-friendly formulation developed by incorporating the drug in a semi-rigid structure. The gels are sticky, simply spreadable with smart esthetic value [6]. The non-profit organization like the International Aloe Science Council has set standards for Aloe vera approval and seal of quality for aloe merchandise with established medicine beneficial [7]. The part of the plant has the cluster of specialized cells called the pericyclic tubules that occur simply at a lower place of the outer inexperienced ring of the leaf. These cells manufacture exudates that incorporate bitter yellow latex with powerful laxative like action [8]. Various aspect effects or toxicity of artificial medication will be overcome by the use of seasoning drugs within the type of appropriate drug delivery system this is often higher patient compatible with less side effect [9]. The aim of the study is that the formulation of aloe containing different types of pharmaceutical dosage forms and its preparation procedure. However, there is an approach to supply the formulations for business production with various kinds of dose form like a gel, hydrogel, emulgel, transdermal patch, beads, tablet, capsule, nanoparticles, microparticles, nanosuspension, cream, face mask, and chewing gum along with environmentally friendly attributes.

9.2 DIFFERENT TYPES OF PHARMACEUTICAL FORMULATIONS PREPARATIONS PROCEDURE/METHOD OF HERBAL FORMULATIONS OF ALOE VERA

9.2.1 TOOTH GEL

Gels are outlined because the semi-rigid systems within which the movement of the phase is restricted by associate degree interlocking three-dimensional network of particles or solvated macromolecules of the phase. The word gel comes from "*gelatin*" and each gel and jelly will be drawn back to the Latin "*gelu*" for "*frost*" and "*gel*" for "*freeze or solidify.*" Gel made of Aloe plant has in style within the market and simple to use as Ayurvedic medicines and cosmetics. The gel of the Aloe plant contains 75 nutrients and to prove an honest favorer remedy. The chemical compound is an oversized category of material consisting of many little molecules that will be cross-connected along to create long chains or to form a gel. Humans have taken advantage of the skillfulness of polymer within the variety of gums, oils, resins, tar, etc. Semi-synthetic polymer like carbopol, HPMC, PVP, etc., will be accustomed to turning out thick formulations with attributes like medium forming, water keeping property, etc. [10, 11].

9.2.1.1   PREPARATION OF TOOTH GEL

Carbapol 940 and Na-CMC were spread in 50 milliliters of water with continuous stirring exploitation mechanical stirrer. Five milliliters of water mixed with the needed amount of Na-benzoate then heated on the water tub to dissolve poorly. The solution was cooled and polyethylene glycol 4000 was added and mixed with 1st solution. Then the needed amount of Aloe vera leaves extract were mixed to the higher than mixture and volume was compared with remaining water. Finally, the full mixed ingredient was mixed to Carbopol 940 gel properly with continuous stirring, and triethanolamine (TEA) were added dropwise formulation for adjustment of needed pH and to get the gel in required consistency [12].

9.2.2   HYDROGEL

Hydrogels are the compound networks with 3D-configuration proficient of consumption of high quantity of water or biological fluids. Their affinity to soak up water is featured due to the presence of hydrophilic clusters like – OH, –COOH, –CONH$_2$, and –SO$_3$H in polymers forming gel structures. The gel contains the character of the binary compound atmosphere and chemical compound composition which may be a cluster of compound materials represented in the formulation. The hydrophilic structure renders them the capability of holding outsized amounts of water in their three-dimensional networks. In-depth, employment of that merchandise during a range of business and environmental areas of application is taken into account of prime importance. Evidently, natural hydrogels were bit replaced by artificial sorts because of their higher water absorption capability, long service life, and wide verities of raw chemical resources [13, 14].

9.2.2.1   PREPARATION OF HYDROGEL

For the preparation of hydrogel, ascorbic acid, Aloe vera gel powder, gelatin, and starch were weighed. The melting point of gelatin and starch were noted. Both were separately dissolved in hot water and kept in a hot water bath until a transparent solution was obtained. The two solutions were mixed with constant stirring and were cooled to room temperature. To this, the ascorbic acid solution along with Aloe vera gel powder was added with constant stirring. The setup was then kept on a rotary shaker for uniform mixing. To this, emulsion polyvinyl pyrrolidone solution was added and thus hydrogel was formed [15].

9.2.3   EMULGEL

Emulgel could be a topical drug delivery system that will be outlined as an instantaneous result of drug-containing medication to the skin to induce the effect of a drug or to forestall disorders. The major disadvantages of gel are the delivery of hydrophobic drugs. This can be overcome by emulgels. Recently, emulgel are used as a topical delivery system containing twin unharnesses system, i.e., a mix of gel emulsion. Since hydrophobic medicine isn't soluble in gel bases, it causes downside throughout the discharge of drugs. It helps the emulgel formulation in hydrophobic medicine incorporation into the oil section, so oily globules spread in the liquid phase leading to o/w emulsion which may be mixed into the gel base. Due to several properties, emulgel could be of higher use as topical drug delivery systems over alternative systems [16].

9.2.3.1   PREPARATION OF EMULGEL

The gel bases were ready by dispersing Carbopol 934 and Carbopol 940 in water one by one with constant stirring at a moderate speed victimization mechanical shaker. Formulation F1, F2, and F3 were ready by Carbopol 934 and F4, F5, and F6 by Carbopol 940 gelling agent. In formulations F7, F8, and F9, the gels were made ready by dispersing HPMC K100 in the heated water at 80°C, and then the dispersion was cooled, and left long for the association of gel. The pH of all the formulations was adjusted to 5.5–6 victimization of TEA. The oily part of the emulsion was ready by dissolving Span-20 oil. Silicon dioxide, being hydrophobic was conjointly dissolved in oil part whereas the liquid phase was ready by dissolving 20–20 in refined water. Methylparaben and propylparaben were dissolved in humectant and mixed with the liquid part. Each of the oily liquid phases was then one by one heated to 70–80°C. Then the oily part was mixed with gel in 1:1 magnitude relation with stirring to get the emulgel [17].

9.2.4 TRANSEMULGEL

Topical drug delivery is also outlined because of the application of a drug-containing formulation to the skin to treat cutaneal disorder directly. The most application of the topical delivery system is bypass first-pass metabolism. Rejection of the risks and inconveniences of blood vessel medical aid due to the various conditions of absorption just like the presence of the catalyst, internal organ evacuation time, etc., are a bit overcome. Transemulgel is employed to stop aches and pains caused by colds, headaches, muscle aches, and injuries. Transemulgel is one amongst the recent technology in NDDS used locally that have characteristics of twin management unharnessed, i.e., an emulsion in combined type, the dose type, etc. [18].

9.2.4.1 PREPARATION OF TRANSEMULGEL

Initially, o/w emulsion containing 1% weight by weight aceclofenac were ready by dissolving a mixture of Span-20 (0.5 ml) in light liquid paraffin (4.5 ml) which acted as oil phase and then the aqueous phase was prepared by dissolving 0.5% tween-20 in distilled water. The two phases of the emulsion were heated separately to 70–80°C, followed by mixing of the two with continuous stirring until the product cooled to room temperature. The above results in the formation of o/w emulsion. The oil in water emulsion was additional beneath stirring, to the ready Aloe vera gel in a 1:1 weight ratio to produce the homogenous product. 0.5 N NaOH was added to obtain the required consistency. This prepared trans-emulgel was stored in an airtight container [19].

9.2.5 NANOPARTICLES

Nanoparticles could also be outlined as particulate dispersion or solid particles with size within the range of 10–1000 nm. In these methods, the drug is entrapped, dissolved, encapsulated, or connected to a nanoparticle matrix.

Relying upon the strategies of preparation, nanoparticles, nanocapsules, nanospheres, etc., may be obtained. Nanoparticles are a system within which the drug is confined to a cavity encircled by a singular chemical compound membrane. The nanoparticle may be a matrix system within which the drug is physically and uniformly spread. In recent year, researches for perishable chemical compound nanoparticles are flourishing. Notably, those coated with a deliquescent chemical compound like poly(ethylene glycol) referred to as long-circulating particles, and also the ability to flow into for a prolong amount time target a specific organ, as a carrier of deoxyribonucleic acid in factor medical care, and their ability to deliver proteins, peptides, and genes [20].

9.2.5.1   PREPARATION OF NANOPARTICLES

The silver nanoparticles (12 nm size) were ready by chemical reduction of the associate degree solution of $AgNO_3$. About 50 metric capacity unit of this solution is supplemental to 30 metric capacity unit of either liquid or ethanolic aloe extract. The entire reaction was dispensed in the presence of air and a constant neutral pH scale. The mixture was smartly stirred at a temperature of 57°C throughout three hours and so heated 2°C/min to achieve 80°C holding for two hours till getting a transparent solution with little suspended particles that might be removed by straightforward filtration (0.45 μm) [21].

9.2.6   MICROPARTICLES

Microparticles are also outlined as a particulate dispersion or solid particles with size within the range of 1 μm–1000 μm. In these systems, the drug is entrapped, dissolved, encapsulated, or connected to a microparticles matrix. Relying upon the strategy of preparation, microparticles, microcapsules, and microspheres are obtained. Microparticles are the compound entities within which the drug is physically and uniformly spread. Microparticulate drug delivery system is the one in every of the simplest processes to supply the sustained and controlled delivery of drugs to provide long periods of your time. The system contains little particles of solid or small droplets of liquids enclosed by walls of natural and artificial compound films of variable thickness and permeableness that acts as a unleash rate dominant substance [22].

9.2.6.1   PREPARATION OF MICROPARTICLES

The methodology used to prepare Aloe vera/chitosan/vitamin E micropar-ticles is as follows: Firstly, 1 g of chitosan was dissolved in 10 ml acetic acid solution (1%, v/v). Aloe vera (0.008 g) and vitamin E oil (0.72 g) was then added beneath stirring (600 rpm) till total solubilization. This solution was atomized in a spray drier MSD 1.0 under following conditions such as outlet air temperature $78 \pm 3°C$, inlet air temperature $120 \pm 2°C$, suspension feed flow rate: 0.30 l/h, aspirator setting: 10, and airflow rate: 500 N l/h [23].

9.2.7   BEADS

Conventional oral dose kind like the pill, capsules, etc., offers a selected drug concentration in circulation that doesn't unleash at the constant rate for a prolonged amount of your time. However, the controlled unleash drug delivery system provides drug release at a pre-controlled, inevitable rate either systemically or domestically for the supposed length of time and opti-mizes the therapeutic result of a drug by dominant its unleash into the body with lower and fewer frequent dosing. Beads are the various microcapsule that employment because the solid substrate on that the drug is coated or encapsulated within the core or coating of beads. Beads may be providing controlled unleash properties. The developed beads may be employed to increase the bioavailability of the drugs [24].

9.2.7.1   PREPARATION OF BEADS

Aloe gel was collected by creating cut any middle of the leave. In 300 mg of sodium alginate, a solution in 10 ml of purified water was added. A 2 g of the gel was mixed at a 1000 rpm using victimization of magnetic stirrer for 15 min. This was then extruded via syringe (no-18) into a 5% $CaCl_2$ solution with gentle agitation at 37°C temperature. The formed beads were allowed to stand for a few minutes in the solution and finally separated by filtration through Whatman paper (size 0.45 mm) [25].

9.2.8   HYDROGEL FILM

In the previous study, PGLD hydrogel films with the linear and hyperbranched structure were prepared via crosslinking reaction with glutaraldehyde and various dicarboxylic acids, and their swelling behavior and mechanical properties. Hydrogel films of hyperbranched PGLD have been made by using the oligomeric crosslinking agent such as carboxyl-terminated poly(ethylene glycol) PEG, with the vision of improving the tensile strength and swelling characteristics of hyperbranched PGLD hydrogel films crosslinked with PEG of dissimilar molecular weights [26].

9.2.8.1   PREPARATION OF HYDROGEL FILM

The solution of Aloe vera (1.0% w/v) and sodium alginate (1.5% w/v) were primarily organized by dissolving in distilled water. Glycerol was added at 15% w/w, based on the molecular mass of the alginate for extending the malleability of the formed films. Afterward, the alginate and Aloe vera in varying proportions (v/v) of 100:0, 95:5, 85:15, and 75:25 were formed. Afterward, 25 ml of every mixture was cast into the Petri dishes of size 9.5 cm and left to dry at room temperature and controlled humidity. After drying, the fabricated films were immersed in $CaCl_2$ aqueous solution at 5.0% w/v for 5 minutes to obtain the hydrogel films. Finally, the ensuing films were washed with purified water and dried at room temperature, before use [27].

9.2.9   NANOSUSPENSION

Nanosuspension could also be outlined because of the mixture dispersion of nanosized drug particles stable by surfactants. They will even be outlined as a biphasic system consisting of pure drug particles spread in a liquid vehicle within which the diameter of the suspended particle is a smaller amount than μm in size. Nanosuspension also a freeze-dried or spray-dried and also the nanoparticles of a nanosuspension may also be incorporated in a very solid matrix. Nano may be a Greek word, suggests that dwarf. Nano which suggests it is the issue of 10^{-9} or one billionths. Nanosuspension may be a long length of your time micronization of poorly soluble drugs by mixture mills or jet mills was most well-liked. The particle size distribution ranges from 0.1 μm [28].

9.2.9.1   PREPARATION OF NANOSUSPENSION

In the process of preparation of topical nanosuspension-based gel formulation, a mixture of 0.25% SSD nanosuspension and 1% AV-gel was incorporated into 0.5% Carbomer 940 gel base and 0.1% sodium salt of methylparaben (as a preservative). All the ingredients were mixed together under continuous stirring with a glass rod to achieve the required viscosity. An aqueous dispersion of 0.25% SSD nanosuspension was added gradually to the ensuing combination with continuous stirring, followed by neutralization of the pH of developed nanogel using a drop of 0.3% ammonia solution [29].

9.2.10   TOPICAL CREAM

The cream is the topical preparations and to be a crucial part of cosmetic merchandise. Cream could also be imperative pharmaceutical merchandise as even cosmetic creams are supported techniques developed by the pharmacy. The unmedicated creams are extremely used in an exceedingly kind of skin condition. Creams were merely ready by a mixture of two or additional ingredients in water, the solvent. Recently with the advancement of technology, newer ways are used for the formulation of creams are available. These sorts of semi-solid preparations are safe to use by the general public and society [30].

9.2.10.1   PREPARATION OF TOPICAL CREAM

Phase inversion temperature technique was utilized to formulate the AVC with the excipients. During this study, w/o emulsion was ready by adding the aqueous phase to the oily phase below continuous agitation. The oily phase consists of light liquid paraffin, white petroleum jelly, olive oil, paraffin wax, cetyl alcohol, stearyl alcohol, and glyceryl monostearate. The elements of the water phase were sodium lauryl sulfate, Aloe vera sap, and distilled water. The ingredients of the oil phase (Phase A) and the aqueous phase (Phase B) mixed slowly to the oil phase and each of the phases was mixed completely with constant stirring to make sure correct emulsification. Once 40–45 min, heating was stopped and once the system earned a temperature of 40°C, a number of drops of pleasant fragrance and eventually a white-colored swish AVC were obtained. The cream formulations were optimized on the premise of organoleptic characteristics and textural parameters by varied the concentration of the emulsifiers. The optimized formulation was kept in properly labeled folded tubes and unbroken at room temperature for more evaluation [31].

9.2.11   FACE MASK

The formulation of the face mask is an important skin health aspect of aesthetics. Facial mask or cosmetics remain the skin clammy and get rid of sebum from the skin surface to preserve appropriate skin composition. The use of appropriate cosmetic, according to the facial skin type, results in healthy skin. Facial formulations are the foremost life cosmetic product utilized for skin rejuvenation. Facial masks are divided into four groups: sheet mask, peel-off masks, rinse-off masks, and hydrogels [32].

9.2.11.1   PREPARATION OF FACE MASK

The burn plant mask formulation is completed with the amount of Aloe vera extract. The mask composition consists of polyvinyl pyrrolidone K-30 (7.55%), BHT (0.12%), polyvinyl alcohol (1.51%), propylparaben (0.10%), methylparaben (0.10%), and water (90.61%). All the materials were weighed and polyvinyl alcohol was coupled with distilled water (6 times), further heated, and stirred until the color is homogeneous. Polyvinyl pyrrolidone K-30 was stirred in the mortar with the addition of a small amount of distilled water. Both the masses were mixed and stirred until homogeneous content was obtained. The mask preparation was mixed with the Aloe vera gel extracts in the base of the stirring mask. The total Aloe vera C content, drying time, and favorite test were estimated [33].

9.2.12   CHEWABLE TABLETS

Chewable tablets are oral indistinct components that are easy to chew and simple to use as self-medication. Recent stress on pediatric drug development made the supply of safe are simple to use and indefinite-quantity formulations are imperative for clinical observations. Tender tablets are widely used in pediatric components that are indefinite quantity kind. Tender tablets confirm the protection of pediatric people [34].

9.2.12.1   PREPARATION OF CHEWABLE TABLET

The fine powder of ethanol extracts of fresh Aloe vera was washed with water, then the kernel of Aloe vera was cut into small pieces and dried on the cupboard with a temperature of 50°C for two days. The contents were dried blended and then immediately sieved in order to get a homogeneous powder. The dried powders were extracted with 96% ethanol and stirred employing a maceration methodology with an electrical stirrer, so the maceration method is shortened to 6–24 hours. In this method, the pulp was soaked once more with 96% ethanol and collected as macerate. Subsequently, ethanol was evaporated by heating at 70°C in a water bath until a thick extract was obtained. To ensure anthraquinone glycosides in the extracts, TLC was employed. Thick Aloe vera extract was made into dried Aerosil 102 pH to obtain a dry powder and then sieved with 60 mesh sieves to obtain a uniform mass [35] (Figure 9.1).

9.2.13 FAST DISSOLVING TABLETS

Solid dose forms are the foremost fashionable owing to low price, simple administration, correct dose self-medication, pain dodging, and also the most significantly the patient compliance. The foremost fashionable solid dose forms are being tablets and capsules. The quick-dissolving drug delivery system begins gaining quality and the acceptance as a brand new drug delivery system and as a result of they are simple to administer and to the higher patient compliance. A quick-dissolving drug delivery system is obtained by the varied techniques that are direct compression, pill molding, freeze-drying, and spray drying nanonization [36].

9.2.13.1 PREPARATION OF FAST DISSOLVING TABLETS

A preliminary screening of the disintegrants like sodium starch glycolate (SSG), crospovidone (CLP), croscarmellose sodium (CCS), and microcrystal-line cellulose (MCC) was conducted. Mannitol was incorporated as a soluble filler to enhance palatableness, impart a cooling sensation and sweet style upon dissolution. Granulation was done out by the dry granulation technique. All the ingredients were compressed and slugs of 0.8 g were turned out at a compression force of 22 ± 1 KN victimization flat two-faced tooling 17 mm in diameter on a one-punch tablet machine. The slugs were then polished and therefore the ensuing granules sieved through sieve no. 20 USP. The granules were further mixed with a glidant lubricant blend containing magnesium stearate (1% w/w) and talc (2% w/w). The granules were compressed using a single punch tablet machine fitted with 8 mm round standard concave punches to produce tablets of thickness 4.56 ± 0.06 mm [37].

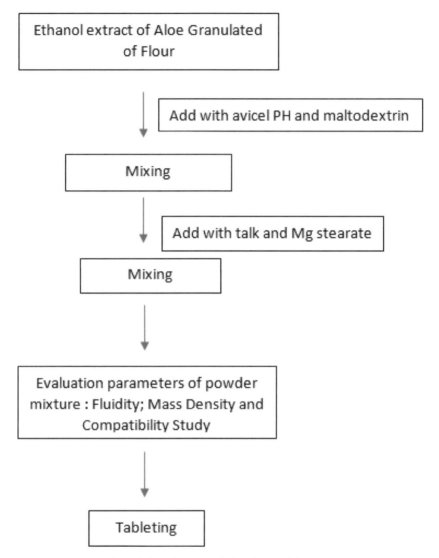

FIGURE 9.1   Preparation of ethanol extract of aloe chews tablet.

9.2.14   CHEWING GUM

Chewing gums are referred to as mobile drug delivery systems. It is administered either regionally or systemically or orally. The medicated chew gum has through the years gained increasing acceptance as a drug delivery system.

Many active ingredients are currently incorporated in medicated chew gum, e.g., halide for prevention of cavity, anti-septic as a native disinfectant, vaso-constrictor for smoking termination, and alkaloid as a keep alert preparation. Additionally, the most range of chew gum is supposed for the hindrance of carries. Recent with the improved technology, an extended grasp alteration, vitamin, and mineral supplementation are presently available [38].

9.2.14.1   PREPARATION OF CHEWING GUM

Aloe vera chewing gum was developed employing a mixture of glycerin, gum bases, liquid aldohexose, sugar, sweeteners (maltitol, xylitol, and aspartame) and flavors like peppermint, eucalyptus, banana, cola, cherries, and cinnamon. In the formulation of Aloe plant-based chewing gum, 10% of Aloe vera dried extract is made to enter into the gum bases. The gum base mixture is softened at a temperature of 70°C. Aloe plant powder, glycerin, liquid glucose, sugar, and other ingredients are added to the base. Finally, at the temperature of 40°C, flavors were added and the mixture of chewing gum was cut into pieces of appropriate sizes [39].

9.2.15   TRANSDERMAL PATCH

The transdermal patch could also be outlined as a medicated adhesive patch that is placed on the skin to deliver a selected dose of medication through the skin and into the blood. It promotes healing to the hurt space of the body. The patch is a bonus of transcutaneous drug delivery route over alteration styles of a delivery system like oral, topical, i.v., i.m., etc. The patch provides a controlled unleash of the medication into the patient, typically through either a porous membrane covering a pool of medication or throughout the body the heat melting skinny layer of medication entrenched within the adhesive [40].

9.2.15.1   PREPARATION OF TRANSDERMAL PATCH

Aloe plant colloidal film was developed from either contemporary aloe or its spray-dried gel powder. This powder was dissolved in de-ionized water and combined with varied concentrations of softener (10%, 15%, 20%, and 30% of glycerin), propylene glycol or polyethylene glycol 400 (PEG400), and co-polymer (5, 10, 15, and 20% of gelatin B, sodium alginate or hydroxypropyl methylcellulose (HPMC) grade E5). The air bubbles within the mixtures were eradicated victimization ultrasonic bath (Crest CP 1100, USA). The mixture was then gently transferred to a Petri dish that was sealed with tin foil, and dried by hot air oven at $50 \pm 2°C$ for night long to make a film or a skin patch. The film was kept in desiccators at room temperature till evaluated [41].

9.2.16   NANOCAPSULE

Nanocapsule is a nanoparticle that is spherical in shape, hollow structure with a diameter ranging from 10 nm to 1000 nm. Nanocapsule is a nano-vesicular system that displays a characteristic core-shell organization in which is confined to a cavity or a reservoir surrounded by a polymer membrane or coating. The drug candidates having poor water solubility have an effect on the bioavailability and therefore the therapeutic index considerably. Nanocapsule drug delivery systems are nice potential for enhancing the bioavailability of poorly water-soluble medicine and to extend the solubility, stability, and permeation of drugs [42].

9.2.16.1   PREPARATION OF NANOCAPSULE

Synthesis of polyamide nanocapsules containing extract in the absence of SC monomer, surfactant, olive oil, and Aloe vera extract was dissolved in 2 ml of acetone. After that, the organic solution was made and drop-by-drop added to 10 ml of water containing the DETA monomer at room temperature while the mixture was being stirred. After ultrasonication, the dilution with water deionized was removed by a rotary evaporator. The final sample was filtered and it was centrifuged which separates the stacked mass without contributing to the extraction of nanoparticles. The nanocapsules were left for about 8 hr at 70°C and then transferred to a freeze drier system to release the water. The final products were brown powder nanocapsules containing Aloe vera extract [43].

9.3   CONCLUSION

Herbal formulations of the pharmaceutical dosage form are additional acceptable and they are safer with minimum facet result than artificial preparations. Natural remedies are the foremost necessary with benefits such as eco-friendliness; less value effective, cut back facet result and their use in medicative formulations for the treatment of burn, wound, healing, inflammation, and also the role of the system in skin health and the burn plant got completely different activities like antifungal, anti-microbial, anti-diabetic, anti-neoplastic, etc. From the variability analysis on completely different pharmaceutical formulations, it is found to be safer and higher by victimization natural plants. Natural remedies are antecedently and recently obtaining additional attention thanks to their potential to treat most diseases. However, many issues like poor solubility, poor bioavailability, low oral absorption, instability, and unpredictable toxicity of natural medicines limit their use. To beat such issues, differing types of pharmaceutical dose kind will play a significant role. Hence, completely different pharmaceutical formulations together with nanoparticles, microparticles, beads, topical cream/gel, tablets, capsules, etc., showed potential to deliver seasoning medicines with higher treatment.

KEYWORDS

- *Aloe barbadensis*
- **Aloe vera**
- **herbal**
- **microcrystalline cellulose**
- **pharmaceutical**
- **sodium starch glycolate**

REFERENCES

1. Thambe, R., Kulkarni, M., Joice, A., & Gilani, I., (2009). Formulation and evaluation of *Aloe Vera* gels. *J. Pharm. Res., 2*(10), 1588–1590.
2. Devi, D. L., Srinivas, B., & Rao, B. N., (2012). An evaluation antimicrobial activity of *Aloe barbadensis* Miller (*Aloe vera*) gel extract. *J. Pharm. Biomed. Sci., 21*(03), 1–4.
3. Katiyar, A., Prajapati, S. K., Akhtar, A., & Vishwakarma, S. K., (2012). *Int. Res. J. Pharm., 3*(10), 143–148.
4. Carac, G., & Gird, C. E., (2016). Antioxidant and antimicrobial potential of extract from *Aloe vera* leaves. *Rev. Chim (Burcharest)., 4*(67), 654–658.
5. Telrandhe, R., (2016). Nanotechnology for cancer therapy: Recent developments. *Eur. J. Pharm. Med. Res., 3*(11), 284–294.

6. Chalke, T., Sharma, K., Nagare, S. K., & Jirge, (2016). Formulation and evaluation of punica topical gel for its content of gallic acid and anti-microbial study. *Int. J. Drug Delivery Tech., 6*(3), 75–78.

7. George, D., Bhat, S., & Antony, B., (2009). Comparative evaluation of the antimicrobial efficacy of *Aloe Vera* tooth gel and two popular commercial toothpastes: An *in vitro* study. *Dental Material Gen Dentistry*, 238–241.

8. Khare, C. P., (2004). *Encyclopedia of Indian Medicinal Plants: Rational Western Therapy, Ayurvedic and Other Traditional Usage, Botany.* Berlin: Springer Science.

9. Partha, N., Snigdha, P., & Laxmidhar, M., (2016). Formulation development and *in vitro* evaluation of dental gel containing ethanol extract of *Tephrosia purpurea* Linn. *Int. J. Pharm. Sci., 8*(8), 132–141.

10. Rathod, H., & Mehta, P. D., (2015). A review on pharmaceutical gel. *Int. J. Pharm. Sci., 1*(1), 33–47.

11. Tambe, R., Kulkarni, M., Joice, A., & Gilani, I., (2009). Formulation and evaluation of *Aloe vera* gels. *J. Pharm. Res., 2*(10), 1588–1590.

12. Shende, V., & Telrandhe, R., (2017). Formulation and evaluation of tooth gel from *Aloe vera* leaves extract. *Int. J. Pharm. and Drug Analysis, 5*(10), 394–398.

13. Ahmed, M. E., (2015). Hydrogel: Preparation, characterization, and application: A review. *J. Adv. Res., 6*, 105–121.

14. Pulasani, S., Boddu, V. K., & Ch, A. B., (2013). A review article on hydrogels. *Int. J. Res. Pharm. Nano Sci., 2*(5), 548–553.

15. Ojha, K., Shenoy, V., Gupta, S., & Suseem, S. R., (2013). Formulation and evaluation of hydrogel with ascorbic acid using *Aloe Vera* gel powder as a drug carrier. *Innovare. J. Sci., 1*(1), 18–20.

16. Thomas, J., Kuppuswamy, S., Sahib, A. A., Benedict, A., & George, E., (2017). A review on emulgel as a current trend in topical drug delivery system. *Int. J. Pharm. Pharm. Res., 9*(3), 273–281.

17. Sumathi, A., & Suriyaprakash, T. N. K., (2016). Formulation and characterization of aloe emulgel using rice hulls as an excipient. *Int. J. Res. Pharm. Pharm., 1*(10), 24–31.

18. Yadav, K. S., Mishra, K. M., Tiwari, A., & Shukla, A., (2016). Emulgel: A new approach for enhanced topical drug delivery. *Int. J. Curr. Pharm. Res., 9*(1), 15–19.

19. Prasanna, R. Y., Haritha, K., Rao, P. S., Vandana, K. R., & Thushara, B. D., (2015). *Current Drug Delivery, 12*(0), 2–7.

20. Mohanraj, V. J., & Chen, Y., (2006). Nanoparaticles: A review. *Trop. J. Pharm. Res., 5*(1), 561–573.

21. Velez, E., Campillo, G., Morales, G., Hincapie, C., Osorio, J., & Arnache, O., (2018). Silver nanoparticles obtained by aqueous or ethanolic *Aloe Vera* extract: An assessment of the antibacterial activity and mercury removal capability. *J. Nanomaterial, 2*(19), 1–7.

22. Kumari, S., Nagpal, M., Aggarwal, G., Puneet, Jain, K. U., & Sharma, P., (2016). Micro particulate drug delivery system: A review. *World J. Pharm. Pharm. Sci., 5*(3), 543–566.

23. Pereira, G. G., Oliveira, S. R., Albernaz, S. M., Canema, D., Weismuller, G., Barros, B. E., Magalhaes, L., et al., (2013). Micro particles of *Aloe Vera*/vitamin E/chitosan: Microscopic, a nuclear imaging and a test analysis for burn treatment. *Eur. J. Pharm. Bioph., 11*(12), 1–7.

24. Kumar, R., Gautam, K. P., & Chandra, A., (2018). Formulation and evaluation of multiple unit floating beads of antiulcer drug. *Asi. J. Pharm., 12*(2), 680–690.

25. Bal, A., Ara, T., Deva, S. A., Madan, J., & Sharma, S., (2013). Preparation and evaluation of *Aloe Vera* gel beads. *J. Global Biosc., 2*(6), 206–216.

26. Kim, S. B., IM, S. J., Bake, T. S., Lee, O. J., Sigeta, M., & Yoshinga, K., (2006). Synthesis of polyglycidol hydrogel film cross linked with carboxyl-terminated poly(ethylene glycol). *Poly J., 38*(4), 335–342.

27. Pereira, R., Mendes, A., & Bartolo, P., (2013). Alginate/Aloe vera hydro gel films for biomedical applications. *Sci. Direct*, 210–215.

28. Kumari, K. V. P., & Rao, S. Y., (2017). Nanosuspension: A review. *Int. J. Pharm., 7*(2), 77–89.

29. Barkat, A. M., Harshita, Ahmad, I., Ali, R., Singh, P. S., Pottoo, H. F., Beg, S., & Ahmad, J. F., (2017). Nanosuspension-based *Aloe Vera* gel of silver sulfadiazine with improved wound healing activity. *Ame. Asso. Pharm. Sci.*, 1–7.

30. Sahu, T., Patel, T., Sahu, S., & Gidwani, B., (2016). Skin cream as topical drug delivery system: A review. *J. Pharm. Biol. Sci., 4*(5), 149–154.

31. Yadav, P. N., Shrivastava, S., Sinha, P., Chanda, D., Luqman, S., & Tandon, S., (2014). Development and evaluation of *Aloe Vera* (L.) burm based topical cream formulation. *Ann. Phyto., 3*(2), 6–65.

32. Nilforoushzadeh, A. M., Amirkhani, A. M., Zarrintaj, P., Moghaddm, S. A., Mehrabi, T., Alavi, S., & Sisakht, M. M., (2018). Skin care and rejuvenation by cosmeceutical facial mask. *J. Cosmet. Dermatol.*, 1–10.

33. Hendrawati, Y. T., Nugrahani, A. R., Utomo, S., & Ramadhan, I. A., (2018). Formulation process making of *Aloe Vera* mask with variable percentage of *Aloe Vera* gel extract. *IOP Conf. Ser. Mater. Sci. Eng.*, 1–8.

34. Michele, M. T., Knorr, B., Vadas, B. E., & Reiss, F. T., (2002). Safety of chewable tablets for children. *J. Asthma, 39*(5), 391–403.

35. Prasetyo, G. D., Setianto, B. A., & Ikhsanudin, A., (2012). Formulation chewable tablets ethanol extract of *Aloe Vera* (*Aloe vera* L.) with the combination of excipients avicel PH 102-maltodextrin. *Pro. Int. Conf. Drug Dev. Nat. Reso.*, 175–182.

36. Yadav, G., Kapoor, A., & Bhargava, S., (2011). Fast dissolving tablets recent advantages: A review. *Int. J. Pharm. Sci. Res., 3*(3), 728–736.

37. Madan, J., Sharma, A. K., & Singh, R., (2009). Fast dissolving tablet of *Aloe Vera* gel. *Tro. J. Pharm. Res., 8*(1), 63–70.

38. Desai, R. T., Dedakiya, S. A., & Bandhiya, M. H., (2011). Medicated chewing gum: A review. *Pharm. Sci., 1*(1), 111–128.

39. Aslani, A., Ghannadi, A., & Raddanipour, R., (2015). Formulation and evaluation of *Aloe Vera* chewing. *Adv. Biom. Res.*, 1–7.

40. Sharma, N., (2018). A brief review on transdermal patches. *Org. Med. Chem. Int. J., 7*(2), 1–5.

41. Puttarak, P., Pichayakorn, W., & Sripoka, K., (2015). Preparation of centella extract loaded *Aloe Vera* transdermal patches for wound healing purpose. *Adv. Mate. Res., 1060*(10), 54–57.

42. Wykar, M., Salunkhe, S. K., Chavan, J. M., Hundiwale, C. J., Gite, K., & Talke, S., (2018). A review on: Nanocapsules. *World J. Pharm. Pharm. Sci., 7*(7), 295–304.

43. Esmaeili, A., & Ebrahimzadeh, M., (2015). Preparation of polyamide nanocapsules of *Aloe Vera* l. delivery with *in vivo* studies. *Ame. Asso. Pharm. Sci., 16*(2), 243–249.

CHAPTER 10

Sesquiterpenes in *Artemisia* and Development of Drugs from Asteraceae

FRANCISCO TORRENS[1] and GLORIA CASTELLANO[2]

[1]*Institute for Molecular Science, University of Valencia,*
PO Box 22085, E–46071 Valencia, Spain, E-mail: torrens@uv.es

[2]*Department of Experimental Sciences and Mathematics, Faculty of*
Veterinary and Experimental Sciences Valencia Catholic University Saint
Vincent Martyr, Guillem de Castro-94, E–46001 Valencia, Spain

10.1 INTRODUCTION

Setting the scene: the simultaneous quantification of five sesquiterpene components after ultrasound extraction in *Artemisia annua*, ethnopharmacological studies for the development of drugs with special reference to Asteraceae and γ-lactone germacranolide budlein A, a sesquiterpene lactone (STL) from *Viguiera robusta*, which alleviates pain and inflammation in a model of acute gout arthritis in mice. Guo group informed the design, synthesis, and cytotoxicity of dihydroartemisinin-coumarin hybrids *via* click chemistry (CC) [1]. They reported the design, synthesis, cytotoxicity, and mechanism of the hybrids as potential anti-cancer agents [2]. They published the design, synthesis, and mechanism of the hybrids as potential anti-neuroinflammatory agents [3]. It was informed bioactive phenolics of the genus *Artemisia*, chromatographic-mass-spectrometric profile of Siberian species and inhibitory potential *vs.* carbohydrate (CH_2O) hydrolysis enzymes (α-amylase, α-glucosidase) [4]. Artesunate attenuates experimental osteoarthritis (OA), inhibiting bone resorption, and $CD31^{hi}Emcn^{hi}$ vessel formation in subchondral bone [5]. γ-Lactone germacranolide costunolide was reviewed as a bioactive STL with diverse therapeutic potential (*cf.* Figure 10.1) [6].

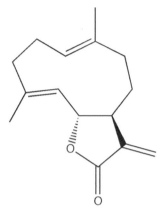

FIGURE 10.1   Chemical structure of costunolide.

The STLs are a large and diverse group of secondary metabolites of many plants, mainly Asteraceae family. Given the therapeutic activity in several experimental models, STLs received attention (e.g., possess anti-inflammatory, analgesic, antitumoral, antimicrobial effects). In terms of rheumatoid arthritis (RA), mice treated with an extract of *Inula helenium* L., containing STLs, ameliorates collagen-induced arthritis reducing NF-κB activation and downstream cytokine production. Budlein A is an STL with antinociceptive and anti-inflammatory properties, related to the inhibition of pro-inflammatory cytokines and neutrophil recruitment. The STLs inhibit NF-κB signaling pathway and diminish inflammatory processes. They are secondary metabolites biosynthesized mainly by species from the plant family Asteraceae. They are present (e.g., lettuce *Lactuca sativa*, chicory *Chicorium intybus* L.) and represent an important part of human diet. They present anti-inflammatory, analgesic, antitumoral, antiparasitic, and antimicrobial activities.

Earlier publications classified 31 STLs [7, 8]. It was informed the tentative mechanism of action, resistance of artemisinin (Art) derivatives [9], reflections, proposed molecular mechanism of bioactivity, resistance [10], chemical, biological screening approaches, phytopharmaceuticals [11], chemical components from *Artemisia austro-yunnanensis*, anti-inflammatory effects, lactones [12], triazole-derived, artesunate, metabolic pathways for Art [13], *A. integrifolia*, *A. capillaris* components and STLs [14]. The aim of this work is to review the simultaneous quantification of five sesquiterpene components after ultrasound extraction in *A. annua*, ethnopharmacological studies for the development of drugs with special reference to Asteraceae, and budlein A, an STL from *V. robusta*, which alleviates pain and inflammation in a model of acute gout arthritis in mice. The purpose of this report is to review recent advances in sesquiterpenes in *A. annua*, ethnopharmacology (EP), drugs related to Asteraceae, and STL budlein A from *V. robusta*, which alleviates pain and inflammation.

10.2 SIMULTANEOUS QUANTIFICATION OF FIVE SESQUITERPENES IN *ARTEMISIA ANNUA*

Simultaneous quantification of five sesquiterpene components was informed after ultrasound extraction in *A. annua* L. by an accurate and rapid ultraperformance liquid chromatography (UPLC)-photodiode array (PDA) assay [15]. The reference standards of δ-lactone Art, γ-lactone arteannuin B (Art B), δ-lactone arteannuin C (Art C), dihydroartemisinic acid (DHAA) and artemisinic acid (AA, *cf.* chemical structures in Figure 10.2) were all isolated from *A. annua*.

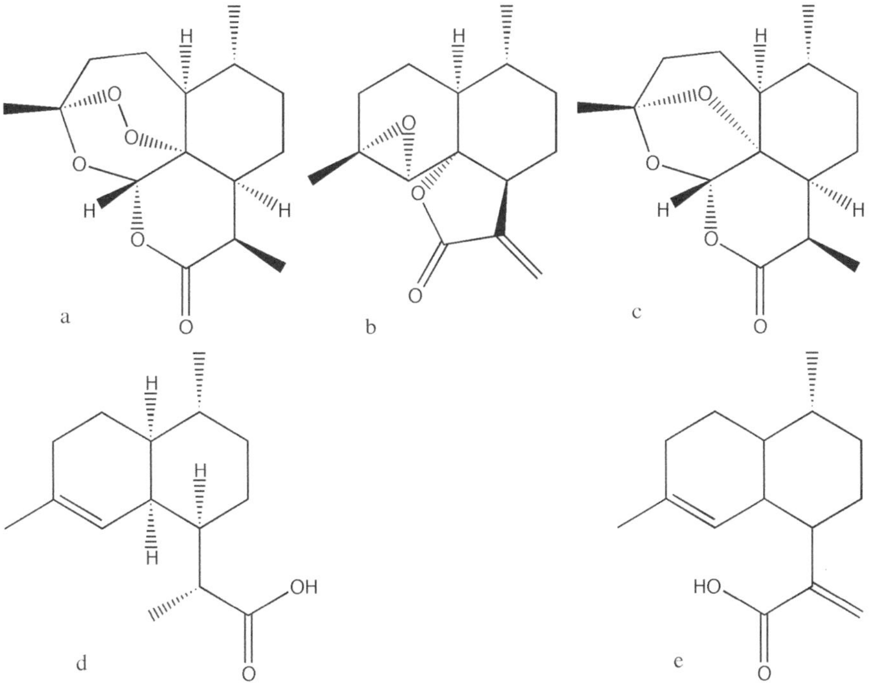

FIGURE 10.2 Structures of marker components in *A. annua*: (a) Art; (b) Art B; (c) Art C; (d) DHAA; (e) AA.

10.3 DEVELOPMENT OF DRUGS WITH SPECIAL REFERENCE TO ASTERACEAE

One of the most important genera of Asteraceae is probably *Artemisia*, as it is used worldwide in traditional medicine (TM), as a source for spices (e.g., *A. dracunculus*), ingredient for liquors (e.g., *A. absinthium*) and as a source for Art [16]. The discovery of Art fundamentally changed the treatment of malaria around the globe and led to the awarding of the Nobel Prize in Physiology or Medicine 2015.

10.4 STL BUDLEIN A ALLEVIATES PAIN AND INFLAMMATION IN A GOUT MODEL

Considering the efficacy of therapies targeting nuclear factor κ-light-chain-enhancer of activated B cells (NF-κB) pathway in RA and the fact that one report exists, addressing the analgesic effect of γ-lactone germacranolide budlein A (*cf.* Figure 10.3), the efficacy of Budlein A extracted from *V. robusta* was evaluated in a model of antigen-induced arthritis (AIA)-induced hyperalgesia and inflammation in mice [17]. Given the pharmacological properties of budlein A, it was investigated the efficacy of budlein A extracted from *V. robusta* (also known as *Aldama robusta*) (Gardner) E. E. Schill and Panero (Asteraceae) in a model of monosodium urate (MSU)-induced gout arthritis in mice [18]. The best analgesic dose of budlein A was addressed.

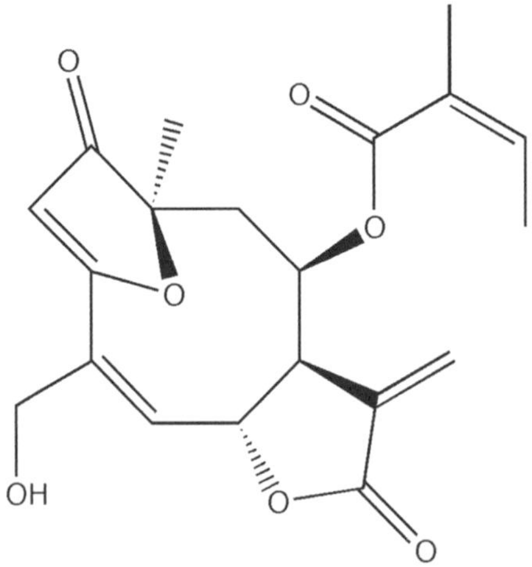

FIGURE 10.3 Representation of the chemical structure of Budlein A.

10.5   DISCUSSION

EP constitutes the scientific basis for the development of therapeutics based on TMs of a number of ethnic groups. The preservation of local knowledge, promotion of indigenous medical systems in primary healthcare and conservation of biodiversity became a concern to all scientists, working at social/ natural sciences interface. Innovations in phytochemical analysis allowed an ever-faster analysis and isolation of bioactive natural products (NPs) and their identification/structure elucidation. Treatment strategies are needed for all diseases, and herbal medicines from TM received attention in the area of prevention or treatment of chronic metabolic disorders, e.g., diabetes mellitus (DM). Many plant species were noted for their anti-DM potential, e.g., plant extracts that inhibited CH_2O hydrolysis enzymes (α-amylase, α-glucosidase), which play a role in CH_2O digestion. The inhibition of the enzymes is a way to avert DM. Asteraceae is a large and widespread family of plants (33 000 spp.). The significance of Asteraceae family for the curative aims was described since centuries. Because of the variety of species in the family, it is important in EP medicine throughout the world. *Artemisia* is a large, diverse genus of plants (480 spp.) belonging to Asteraceae. A rise appeared in research of *Artemisia* phytocomponents with antimalarial, cytotoxic, antihepatotoxic, antibacterial, and antioxidant (AO) activity (AOA). A number of *Artemisia* aqueous and alcoholic extracts possess an anti-DM effect caused by hypoglycaemic action. Pharmacological data evidence that the extracts of *A. absinthium*, *A. afra*, *A. amygdalina*, *A. dracundulus*, *A. judaica*, *A. herba-alba*, *A. ludoviciana*, and *A. sphaerocephala* were effective in streptozotocin- and alloxan-induced hyperglycemia experimental animal models, because of their ability to reduce blood glucose level and protect *vs.* DM-caused metabolic aberrations. The unique reported application of *Artemisia* drugs in humans was conducted on type-2 DM individuals, and used *A. absinthium* capsules for 30d. Blood glucose level was reduced 32% compared to the baseline, which suggests that other *Artemisia* spp. present anti-DM properties.

The STLs constitute a major class of bioactive NPs. One of the naturally occurring STLs is γ-lactone germacranolide costunolide, which was extensively investigated for a wide range of bioactivities. Multiple preclinical studies reported that costunolide possesses AO, anti-inflammatory, antiallergic, bone remodeling, and neuroprotective, hair growth-promoting, anticancer, and anti-DM properties. Many bioactivities are supported by mechanistic details (e.g., modulation of a number of intracellular signaling pathways involved in precipitating tissue inflammation, tumor growth and progression, bone loss, neurodegeneration). The key molecular targets of costunolide include intracellular kinases [e.g., mitogen-activated protein kinases (MAPKs), protein kinase B (PKB, Akt), telomerase, cyclins, cyclin-dependent kinases (Cdks)] and redox-regulated transcription factors [e.g., NF-κB, signal transducer and activator of transcription (STAT), activator protein (AP)-1]. The compound decreased the production and expression of proinflammatory mediators [e.g., *cyclo*-oxygenase (COX)-2, inducible nitric oxide (NO) synthase (iNOS), NO, prostaglandins, cytokines].

Artemisinin was used in Chinese herbal medicine (CHM) for more than 2000 years. The small molecule artesunate, a derivative of Art, was used to treat various kinds of diseases ranging from malaria to RA. The findings broaden the potential application of artesunate, which attenuated subchondral bone deterioration (e.g., suppressing dramatic bone resorption) inhibiting hetero-topic bone formation *via* interrupting transforming growth factor (TGF)-β signaling and abrogating aberrant blood vessel formation in the subchondral bone in early-stage OA. Specifically, articular cartilage degeneration was alleviated, indicating that maintaining stabilization of subchondral bone microarchitecture in early OA was seen as an effective therapeutic approach. Several limitations were present in this study. Compared to oral adminis-tration, intraperitoneal injection is not a considerably more convenient route, especially for those who need to administer repeatedly medications. Although the molecular model indicates the artesunate-receptor activator of NF-κB (RANK) and -TGF-β receptor-I interactions, the effect needs to be further demonstrated by transgenic and knock-out/in models.

10.6   FINAL REMARKS

From the present results and discussion, the following final remarks can be drawn.

1. STLs form a large, structurally diverse group of NPs found almost universally in plants. Investigation of the therapeutic potential of STLs yielded candidates for pharmaceutical development. Costunolide is a well-known STL, which was isolated from a number of plant species. Costunolide possesses antioxidant, anti-inflammatory, antiallergic, bone remodeling, neuroprotective, antimicrobial, hair growth-promoting, anticancer, and antidiabetic properties. Pharmacokinetic studies showed that costunolide is bioavailable. However, most studies were conducted in cultured cells or *via* an *in vitro* system. Considering the therapeutic

value of costunolide, to examine further the effects of costunolide in a number of other animal models results interesting, to reveal the subacute and chronic toxicities, detailed elucidation of molecular mechanisms of action and structural modifications, to develop therapeutics based on costunolide or derivatives.

2. The UPLC method was used to analyze quantitatively compounds (artemisinin, arteannuin B, arteannuin C, DHAA, artemisinic acid) in the extract of *A. annua*. The existing research measures only four components in *A. annua*. It is not easy to grasp accurately the transformation and dynamic equilibrium of sesquiterpenes in *A. annua* biosynthesis. The ultraviolet or evaporative light scattering detector method is not sensitive enough, but UPLC uses a smaller particle size column for better resolution and sharper peaks, which improves sensitivity. The concern is to provide a basis for the mutual transformation and dynamic balance of the sesquiterpenes, in the biosynthesis process, *via* quantitative analysis of *A. annua*.

3. Budlein A reduced pain and inflammation in a model of acute gout arthritis in mice. Molecules with the ability of targeting nuclear factor κ-light-chain-enhancer of activated B cells activation and inflammasome assembly, e.g., budlein A, are approaches to treat gout flares.

ACKNOWLEDGMENTS

The authors thank support from Fundacion Universidad Catolica de Valencia San Vicente Martir (Project No. 2019-217-001UCV).

KEYWORDS

- **anti-inflammatory**
- **antioxidant**
- ***Artemisia annua***
- **costunolide**
- **liquid chromatography**
- **quantitative analysis**
- **sesquiterpene component**

REFERENCES

1. Tian, Y., Liang, Z., Xu, H., Mou, Y., & Guo, C., (2016). Design, synthesis and cytotoxicity of novel dihydroartemisinin-coumarin hybrids *via* click chemistry. *Molecules, 21*, 758–1–15.

2. Yu, H., Hou, Z., Tian, Y., Mou, Y., & Guo, C., (2018). Design, synthesis, cytotoxicity, and mechanism of novel dihydroartemisinin-coumarin hybrids as potential anti-cancer agents. *Eur. J. Med. Chem., 151*, 434–449.

3. Yu, H., Hou, Z., Yang, X., Mou, Y., & Guo, C., (2019). Design, synthesis, and mechanism of dihydroartemisinin-coumarin hybrids as potential anti-neuroinflammatory agents. *Molecules, 24*, 1672–1–20.

4. Olennikov, D. N., Chirikova, N. K., Kashchenko, N. I., Nikolaev, V. M., Kim, S. W., & Vennos, C., (2018). Bioactive phenolics of the genus *Artemisia* (Asteraceae): HPLC-DAD-ESI-TQ-MS/MS profile of the Siberian species and their inhibitory potential against a-amylase and a-glucosidase. *Front. Pharmacol., 9*, 756–1–27.

5. Li, Y., Mu, W., Xu, B., Ren, J., Wahafu, T., Wuermanbieke, S., Ma, H., et al., (2019). Artesunate, an anti-malaria agent, attenuates experimental osteoarthritis by inhibiting bone resorption and CD31[hi]Emcn[hi] vessel formation in subchondral bone: *Front. Pharmacol., 10*, 685–1–13.

6. Kim, D. Y., & Choi, B. Y., (2019). Costunolide: A bioactive sesquiterpene lactone with diverse therapeutic potential. *Int. J. Mol. Sci., 20*, 2926–1–21.

7. Castellano, G., Redondo, L., & Torrens, F., (2017). QSAR of natural sesquiterpene lactones as inhibitors of Myb-dependent gene expression. *Curr. Top. Med. Chem., 17*, 3256–3268.

8. Torrens, F., & Castellano, G. (2020). Structure-activity relationships of cytotoxic lactones as inhibitors and mechanisms of action. *Curr. Drug Discov. Technol. 17*, 166–182

9. Torrens, F., Redondo, L., & Castellano, G., (2017). Artemisinin: Tentative mechanism of action and resistance. *Pharmaceuticals, 10*, 20-4-4.

10. Torrens, F., Redondo, L., & Castellano, G., (2018). Reflections on artemisinin, proposed molecular mechanism of bioactivity and resistance. In: Haghi, A. K., Balköse, D., & Thomas, S., (eds.), *Applied Physical Chemistry with Multidisciplinary Approaches* (pp. 189–215). Apple Academic–CRC Press: Waretown, NJ.

11. Torrens, F., & Castellano, G. (2020). Chemical and biological screening approaches to phytopharmaceuticals. In: Pourhashemi, A., Deka, S. C., & Haghi, A. K., (eds.), *Research Methods and Applications in Chemical and Biological Engineering*. Apple Academic–CRC Press: Waretown, NJ, pp. 3–12

12. Torrens, F., & Castellano, G. (2020). Chemical components from *Artemisia austro-yunnanensis*: Anti-inflammatory effects, and lactones. In: Pogliani, L., Torrens, F., & Haghi, A. K., (eds.), *Molecular Chemistry and Biomolecular Engineering: Integrating Theory and Research with Practice*. Apple Academic–CRC Press: Waretown, NJ, pp. 73–83

13. Torrens, F., & Castellano, G. (2020). Triazole-derived, artesunate and metabolic pathways for artemisinin. In: Shinde, R. S., & Haghi, A. K., (eds.), *Modern Green Chemistry and Heterocyclic Compounds: Molecular Design, Synthesis, and Biological Evaluation*. Apple Academic–CRC Press: Waretown, NJ, pp. 137–144

14. Torrens, F., & Castellano, G.*Artemisia integrifolia*, A. capillaris components and sesquiterpene lactones. In: Yaser, A. Z., Khullar, P., & Haghi, A. K., (eds.), *Green*

Materials and Environmental Chemistry: New Production Technologies, Unique Properties, and Applications. Apple Academic–CRC Press: Waretown, NJ, in press.

15. Ruan, J., Liu, Z., Qiu, F., Shi, H., & Wang, M., (2019). Simultaneous quantification of five sesquiterpene components after ultrasound extraction in *Artemisia annua* L. by an accurate and rapid UPLC-PDA assay. *Molecules, 24*, 1530-1-15.

16. Panda, S. K., Da Silva, L. C. N., Sahal, D., & Leonti, M., (2019). Editorial: Ethno pharmacological studies for the development of drugs with special reference to *Asteraceae*. *Front. Farmacol., 10*, 955–1–2.

17. Zarpelon, A. C., Fattori, V., Souto, F. O., Pinto, L. G., Pinho-Ribeiro, F. A., Ruiz-Miyazawa, K. W., Turato, W. M., et al., (2017). The sesquiterpene lactone, budlein A, inhibits antigen-induced arthritis in mice: Role of NF-kB and cytokines. *Inflammation, 40*, 2020–2032.

18. Fattori, V., Zarpelon, A. C., Staurengo-Ferrari, L., Borghi, S. M., Zaninelli, T. H., Da Costa, F. B., Alves-Filho, J. C., et al., (2018). Budlein A, a sesquiterpene lactone from *Viguiera robusta*, alleviates pain and inflammation in a model of acute gout arthritis in mice. *Front. Pharmacol., 9*, 1076-1-13.

Index